saguaro national monument

ARIZONA

Napier Shelton
*based on an earlier
work by Natt Dodge*

NATURAL HISTORY SERIES

1972
National Park Service
U.S. DEPARTMENT of the INTERIOR

preface

This book is a simple account of the natural history of Saguaro National Monument. It is intended to help you understand the relationships between land, climate, plants, wild animals, and man in the environment of a hot desert. While it includes brief profiles of many representative species, it is not intended to serve as a guide to the monument. It does indicate where the several distinctive natural communities exist, and when and where to look for certain plants and animals. For identification purposes, you will need field guides.

The present edition is my revision of the 1957 book by Natt Dodge, then naturalist for the Southwest Region of the National Park Service. The first five and the last two chapters are essentially new; the central chapters on plants and animals remain largely as written in the first edition.

The authors wish to thank former and present members of the monument staff for their help and companionship, in both field and office. We are particularly grateful to Chief Naturalist Harold T. Coss, Jr., who devoted much time and effort to obtaining many of the photographs and, with Park Biologist Warren F. Steenbergh, gave the manuscript a thorough scrutiny. The cooperation and hospitality of Superintendent Harold R. Jones have created the best possible climate for work on the revised edition. National Park Service geologist Robert H. Rose contributed in very great measure to the geological content. Finally, our thanks go to the many students of desert life on whose knowledge this book has been built, and to monument visitors who ask questions—for their concern gives hope for better relations between man and nature.

— N.S.

contents

the desert scene

Scattered through the wide, lonely Sonoran Desert, isolated mountain ranges raise jagged blue silhouettes against the sky. The high ones wear a crown of dark pines and a speckled mantle of oaks. Lapping against their feet is the desert sea, studded with the green masts of giant saguaro cactuses rising above a motley crew of tough, strange, often handsome subordinates.

On either side of the Santa Cruz Valley in southeastern Arizona, Saguaro National Monument embraces two of these ranges, with the desert lands at their feet. Although it was established primarily to preserve impressive stands of saguaros, it gains wholeness by including the mountains that shed the rocky soil on which the saguaros grow. The monument is in two sections, some 30 miles apart. The *Rincon Mountain Section,* east of Tucson, includes in its 99 square miles a spectrum of plantlife ranging from saguaro communities at the low elevations to a wet fir forest on the north slope of 8,666-foot Mica Mountain. The 24-square-mile *Tucson Mountain Section,* about 12 miles west of the city, has a denser stand of saguaros, on the lower slopes of a wild jumble of volcanic mountains dominated by 4,687-foot Wasson Peak.

Within the cactus forests and upon the mountain slopes, life in myriad forms goes on in a delicate and continuous adjustment to changing environments. Over millennia, mountains rise and crumble, species evolve and fade from the scene. Over days and weeks and years, populations of plants and animals rise and fall, in response to thousands of interactions which we are only beginning to understand. In the following pages, we will meet the main characters in this natural drama (including some of the most improbable in all of nature); we will investigate the environmental stage upon which they perform; and we will try to understand something of the play itself, including our own role in it.

Brittlebush display along Cactus Forest Drive.

1

The Saguaro Forest in the Tucson Mountain Section.

the sonoran desert
and the monument

Saguaro National Monument is at the northeastern edge of the Sonoran Desert. Named for the state of Sonora, Mexico, in which the greater part of it lies, this, one of four major deserts in North America, is distinguished by differences in climate and vegetation. The Great Basin Desert, mainly in Nevada and Utah between the Rockies and the Sierra Nevada, has cold winters, sparse precipitation distributed fairly evenly throughout the year, and rather simple vegetation dominated by the low shrubs, sagebrush and saltbrush. Immediately to the south, in southern Nevada and southeastern California, is the Mohave Desert, with cool winters during which most of the year's precipitation comes, and with plant cover consisting mostly of shrubs such as creosotebush. At higher elevations grows the Joshua-tree, a giant yucca. The Chihuahuan Desert, with cool winters and summer rainfall, covers the broad plateau of north-central Mexico, extending into southern New Mexico and west Texas. Its vegetation consists mosty of small cactuses, spiny shrubs, and succulent-leaved plants such as yuccas.

Lying between the Mohave and Chihuahuan deserts, the Sonoran Desert has both winter and summer rainfall, with spring and autumn droughts. Its mild winters and bi-seasonal rainfall encourage a vegetation far surpassing in lushness and variety that of the other deserts. From west to east, the land rises and precipitation increases. Yuma, in southwestern Arizona, lying at 141 feet above sea level, gets about 3 inches a year. Tucson, in southeastern Arizona, is at 2,400 feet elevation and averages 11 inches a year. In the western part, where plains are extensive and mountain ranges low and far apart, a very few small-leaved species such as creosotebush dominate. Toward the east, mountain ranges become more numerous, shedding material on which more kinds of plants, particularly paloverde, mesquite, and cactuses, assume leading roles in the vegetative cover. Both sections of the monument lie within the eastern, wetter, more diversified part of the Sonoran Desert known as the Arizona Upland. Organ Pipe Cactus National Monument, to the southwest, is in an area transitional between the Arizona upland and Colorado River lowland phases of the Sonoran Desert.

3

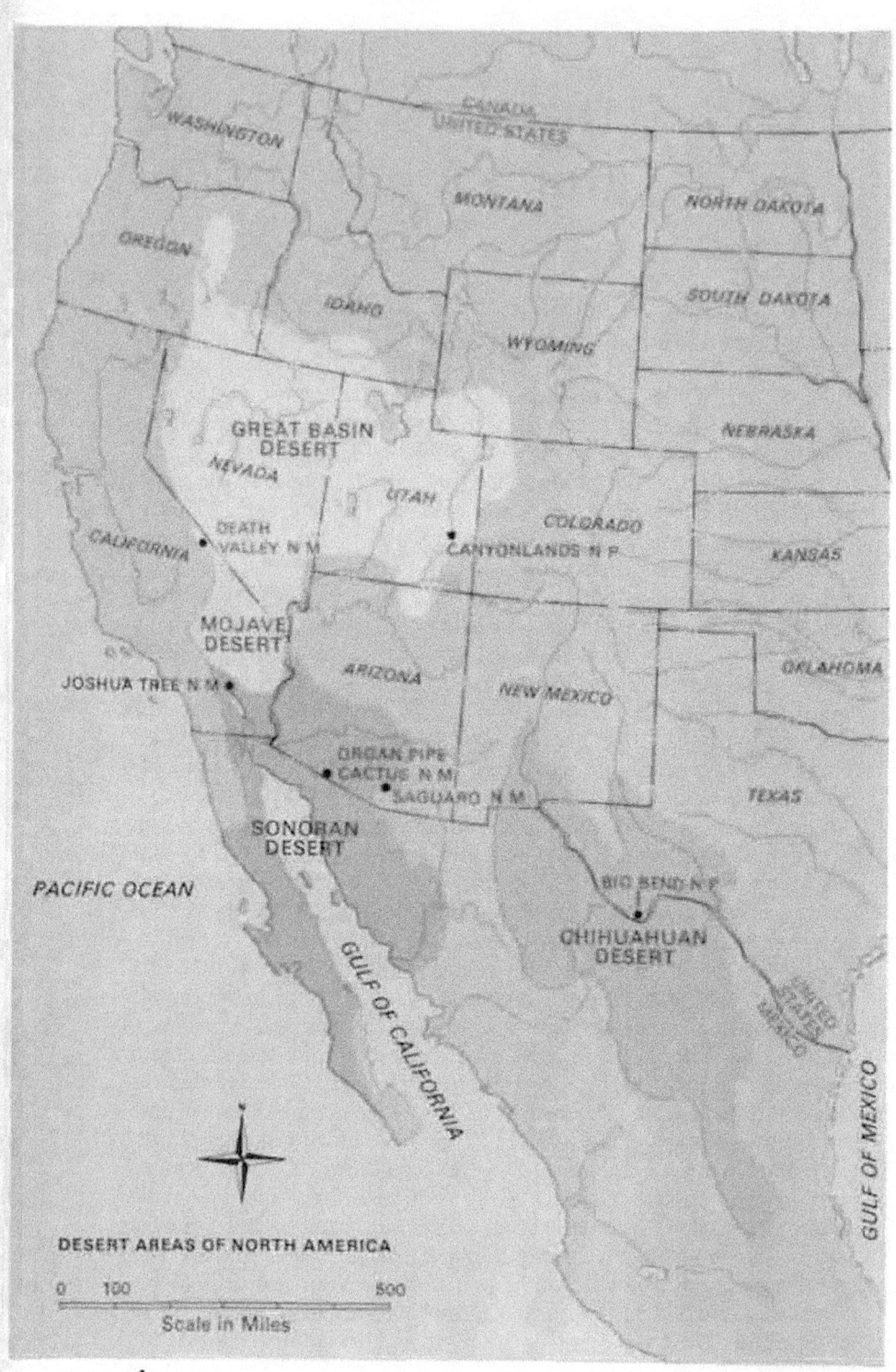

DESERT AREAS OF NORTH AMERICA

0   100   500

Scale in Miles

4

But even between the two sections of Saguaro National Monument there are differences. There is the obvious fact that the Tucson Mountains are much lower and smaller in mass than the Rincons. And there is the not-so-obvious fact that the Tucson Mountain Section extends to lower elevations (2,200 as opposed to 2,700 feet). These conditions are reflected in the somewhat warmer environment of the western section. This helps to explain why certain Sonoran Desert plants and animals here reach their northeastern limits; among them are ironwood, desert iguana, desert horned lizard, western shovel-nosed snake, sidewinder, desert kangaroo rat, and the Le Conte thrasher. The Rincon Mountain Section, by virtue of its higher elevations, has plants and animals—such as ponderosa pine, Steller's jay, and whitetail deer—not found in the western section.

On the geologic time scale the Sonoran Desert, like all the earth's present deserts, is a recent development. Some 50 million years ago tropical forests grew here, as they did over most of southern North America. As the Southwest gradually became drier, new species adapted to the new conditions evolved; and the general type of vegetation changed—with some retrogression during wetter periods—from forest to savanna (grasslands with scattered trees) to arid subtropical scrub (such as is now found in southern Sonora State), and finally to the plant communities of today's Sonoran Desert. The last stage has occurred only during the past few million years.

The assemblage of species from which most of our present desert and lower mountain plants were derived is known as the Madro-Tertiary Flora, to denote its center (Sierra Madre in Mexico) and time (Tertiary period—one million to 65 million years ago) of early development. Reflecting these origins, a majority of the species growing below 6,000 feet elevation in the monument today are basically Mexican or Central or South American in their distribution.

On the higher mountains of the Sonoran Desert, however, exist species with an entirely different ancestry. Most of the plants of the pine and fir forests of these high places, like the plants of the Great Basin Desert, derive from the Arcto-Tertiary Flora, which dominated the northern part of our continent during Tertiary times. As climate changes, one or the other of these great plant assemblages will benefit at the expense of the other. If cooler, wetter periods return, as they did during the Pleistocene epoch, ponderosa pines, now found above 6,000 feet, may again grow nearly as low down as today's cactus forests. But if the present long-term drying trend continues, the northern plants will eventually be squeezed off their mountaintops.

Animals, to some extent, reflect the same dichotomy of origins we have seen in the plants. Thus, on the mountaintops supporting

5

pine and fir forests, a northern contingent of animals predominates, while on the lower slopes and desert the fauna has a Mexican character. The canyons and oak-pine forests of middle elevations in southeastern Arizona are particularly exciting, because in these biological islands, isolated by surrounding desert, live many "Mexican" species of animals found nowhere else in the United States. Naturally, mountain ranges nearest the border—such as the Huachucas and Chiricahuas—have the greatest numbers of Mexican specialties; but the Rincons have their share too.

The wonderful diversity of plant and animal life in Saguaro National Monument depends on a diversity of habitats. These in turn owe their existence to a widely varying triumvirate of environmental factors—climate, soil, and topography. To understand the biological interplay that goes on here, we first must know something of the environmental conditions that circumscribe it.

Plant communities on Mica Mountain.

Geological Time Table

Era	Period	Epoch	Years before present time	Major geological events
Cenozoic	Quaternary	Recent	10,000	Cascade Range and Sierra Nevada uplifted
		Pleistocene	1,000,000	
	Tertiary	Pliocene	12,000,000	
		Miocene	26,000,000	
		Oligocene	38,000,000	
		Eocene	54,000,000	Rocky Mountains formed
		Paleocene	65,000,000	
Mesozoic	Cretaceous		136,000,000	
	Jurassic		195,000,000	
	Triassic		250,000,000	Appalachian folding
Paleozoic	Permian		280,000,000	
	Pennsylvanian		320,000,000	
	Mississippian		345,000,000	
	Devonian		395,000,000	
	Silurian		440,000,000	
	Ordovician		500,000,000	
	Cambrian		570,000,000	
Precambrian				

Summer thunderstorm in the Rincon Mountain Section.
The Catalina Mountains rise in the background.

climate:
the vital factor

Climate is the chief arbiter of life on earth. Each plant and animal, including man, has tolerance limits for heat and water below or above which it dies. The Sonoran desert, with its low and uncertain rainfall and high summer temperatures, thus presents one of earth's most taxing environments. For most desert plants and animals, the main problem presented by this climatic combination is a scarcity of water. Why, we might ask, is this part of the globe so dry?

Essentially, the dryness of the Sonoran and other southwestern deserts results from their geographic situation. First, these lands lie east of a persistent high-pressure cell over the eastern Pacific—part of a belt of high-pressure cells which encircles the earth at this latitude. High-pressure systems in the Northern Hemisphere have descending, warming, dry air in their middles and along their eastern sides. The air that flows eastward from the "East Pacific high" thus contains little moisture, and dictates desert conditions over much of the Southwest. Furthermore, the mountains that more or less ring these deserts intercept much of the meager incoming moisture, by forcing the air to rise and cool so that its moisture condenses and falls to the ground.

But during two periods of the year, the desert does receive rain. In winter, the East Pacific high shifts southward with the sun, reducing its effects on the Southwest. Now the low pressure storm systems, which at this season move eastward across the continent, can sometimes dip far enough south to bring rain to the Sonoran Desert. These rains are usually gentle, and cover large areas. Their clouds may cover the sky for several days at a time.

By April, the East Pacific high has returned far enough northward to resume its control of southwestern weather. Now the fore-summer drought sets in, intensified by gradually rising temperatures. Day after day the sun climbs higher in the sky, baking the earth and testing the endurance of desert life. Then, sometime in late June or early July, clouds begin forming in the afternoons over the mountains; finally, one day they build up to great thunderheads. The drought ends suddenly as rain pours down and lightning cracks the sky. Down the canyons and along the desert washes rages the water, brown with soil. In a short time, the storm

9

is over, and the water disappears into the ground. Just 1 mile away, rain may not have fallen.

These intense, local summer rains, so different from the winter ones, have a different origin as well. They are spawned by the Atlantic's Bermuda high, which has shifted northwestward and sent moist air from the Atlantic and Gulf of Mexico flowing off its western edge toward the deserts. This air rises quickly when it strikes hot, high mountains such as the Rincons, and its moisture condenses to fall as rain.

Some time in September the rains of summer stop, as the Bermuda High moves south again, removing its influence and returning the East Pacific high to a position of dominance. The sun and wind again wring moisture from the soil, but with diminishing effect. The thermometer no longer reaches the 100's as it did in June, July, and August. In November or December, the winter rains begin. Water stays longer in the soil, because temperatures now range between 30° and 80°. Life in the desert begins a new cycle. This rainy period usually lasts into March.

On the mountain slopes, of course, the same seasonal patterns of heat and precipitation occur, but they are tempered or enhanced by elevation. With increasing elevation, temperatures decline and precipitation increases. When July temperatures at headquarters are averaging in the nineties, they will be in the sixties on top of the Rincons. In the cactus forest at the foot of Tanque Verde Ridge only about 10 to 12 inches of rain fall in a year, while Manning Camp, at 8,000 feet, gets about 18 to 25 inches. These up-slope gradients of temperature and precipitation, as we shall see, have pronounced effects on plant and animal life.

Gila monster.

Banded augen gneiss on Cactus Forest Drive.
Javelinas formerly took refuge from the sun in the dens
under these overhanging rocks.

rocks:
foundation and soilmakers

Soils are derived from the fragmentation and decomposition of rocks. Combinations of soils and climate, varying from place to place, create an almost infinite number of environments with differences so subtle and small as to make it appropriate to refer to them as microenvironments. Though these differences be minor, they are often sufficient to create niches each of which becomes a habitat for a particular group of plants and animals. Thus it is that the saguaros and other desert vegetation in Saguaro National Monument and vicinity are found in particular environmental or ecological niches where just the right combinations of soils, moisture, sunlight, temperature and other factors are present.

Since the rocks provide the foundation and the source of soils that support the plant and animal life, it is helpful to understand something about the origin and evolution of the rocks and the landscape of which they are a part. Clues to this story are to be found in the composition of the rocks and their relationships to one another.

Rocks of the three major classes—igneous, metamorphic, and sedimentary—are found within the monument. The igneous rocks include granites and various kinds of lava flows, together with some intrusive dikes and veins. The metamorphic rocks are represented by gneiss and schist. Lastly, sedimentary rocks occur as limestone, sandstone, and alluvial fill material. Each of these great classes of rocks was formed in a different manner, and these differences reveal the nature of the events which are a part of the geologic history of the monument and surrounding area.

Catalina Gneiss (pronounced NICE) is the predominant rock that visitors see in the Rincon Mountains and the Tanque Verde Ridge; exposures of granite and schist are also found. Gneiss is a coarse-grained metamorphic rock resembling granite, having a banded appearance, and consisting of alternating layers of different minerals such as feldspar, quartz, mica, and hornblende. The banding and texture reveal that the gneiss, now exposed by uplift and erosion, was formed from parent rocks deep below the earth's surface during the Precambrian Era more than a half billion years ago. There the parent rocks, subjected to pressure and heat, melted, flowed, and crystallized, before resolidification.

13

Though the gneiss of the Rincon Mountains is ancient, the uplift that raised it to its present height is thought to have occurred rather recently (within the past 24 million years), during the period when most of the Sonoran Desert ranges apparently were formed. In age, these mountains thus would fall somewhere between the older Rockies and the younger Sierra Nevada.

The schist, which underlies much of the Cactus Forest, where it is exposed along washes, was probably formed during the Cretaceous period more than 65 million years ago. Like the gneiss, the schist is a metamorphic rock, having been formed by the transformation of parent rocks at depth under great pressure and heat. Due mainly to differences in the composition of the parent rock, together with its mica content, the rock that resulted was platy and cleavable (accounting for its classification as schist).

Granite, which forms Wasson and Amole peaks in the Tucson Mountain Section, is an igneous rock that (like the metamorphic gneiss and schist) originated at depths below the earth's surface. The granite was formed by the solidification of molten rock material that moved upward en masse from greater depths, rather than by the alteration of ancestral rocks. Erosion of the uplifted land mass has not only stripped away the overburden but has developed valleys deeply incised into the granite itself.

Volcanic rocks in the form of rhyolite, andesite, and basalt flows also are exposed, chiefly in the Tucson Mountain section of the monument. These are all extrusive igneous rocks composed of magma that solidified after reaching the earth's surface through vents or fissures. There are differences in the lava flows, reflecting differences in composition, temperature, and other conditions of the magma from which they were formed. The rhyolite is lighter in color than andesite, and it is somewhat richer in feldspar. Basalt, on the other hand, is dark, is deficient in feldspar and quartz, and contains relatively large amounts of the darker minerals such as hornblende, pyroxene, and olivene. Lava flows of the rhyolite and andesite variety occur in the Cactus Forest locality of the Rincon Mountain Section. Cat Mountain Rhyolite is the name given to a rhyolite flow of Tertiary age that forms the topmost layer in much of the Tucson Mountain Section. A small exposure of basalt is also found here.

Limestone, sandstone, and shale formations are also exposed at various places within the monument. They are reliable indicators of seas that covered the area during one or more times of the geologic past. Comparisons with the limestone formations farther to the south in the Colossal Cave region indicate that the limestone in the monument was formed as far back as late in the Paleozoic Era (345 million years or more ago). This age is suggested by types of fossils, including fragments of crablike trilobites

14

and crinoid (sea lily) stems, found in limestones of the same age near Colossal Cave.

The Recreation Redbeds of the Tucson Mountain Section is one of the important sandstone members exposed in the monument. Geologists believe that its deposition occurred early in the Cretaceous Period more than 65 million years ago. Above the Recreation Redbeds lies a formation consisting mainly of limestone and commonly known as the Tucson Mountain "Chaos"— an appropriate name, in view of the geologists' meager knowledge of it. Within the Chaos formation, believed to be lower to Middle Tertiary in age, are limestone blocks of greater age than the rocks within which they are entombed. No generally accepted explanations have been advanced as to how these relationships developed.

The processes that destroy mountains continue concurrently with those that build them. Temperature changes, weathering, and downslope creep under the influence of gravity are among the agents which destroy the rocks and eventually convert them into soil. Though these processes work rapidly from a geological standpoint, in terms of the average human life span, they progress at an imperceptible rate. Violent thunderstorms and cloudbursts, however, cause a massive, often spectacular movement of boulders, gravel, sand, and silt by torrential streams.

In desert regions, most of the runoff from storms sinks into the slopes, dropping its burden along the way. Big rocks are dropped early as the carrying power of the water diminishes with the speed of flow; smaller fragments travel farther. The finest material is carried far out into the basins between mountain ranges, gradually filling them. (The alluvial material in the Tucson Basin is estimated to be 2,000 to 5,000 feet thick.) Thus desert mountains tend to bury themselves in their own debris.

Alluvial fans (fan-shaped deposits built by rivers flowing from mountains into lowlands) form at the mouths of canyons. Sometimes the alluvial fans of adjacent canyons coalesce, forming the long, sweeping slope known as a bajada (pronounced ba-HA-da). In other places the eroded bedrock extends outward from the bases of desert mountains, forming "pediments," which are usually covered by a veneer of alluvial material. Pediments of this kind stretch from the lower slopes of the Rincon Mountains. In the Tucson Mountains the alluvial material deepens rapidly toward the Avra Valley because the bedrock at the base of the mountain dips steeply downward.

Serious gaps remain in the story of the origin and evolution of the landscape in Saguaro National Monument. Moreover, it is difficult to establish the exact sequence in which the various events occurred. The composition, texture, and relationships of the rocks, however, do reveal much about the nature of the events and the processes that were involved.

15

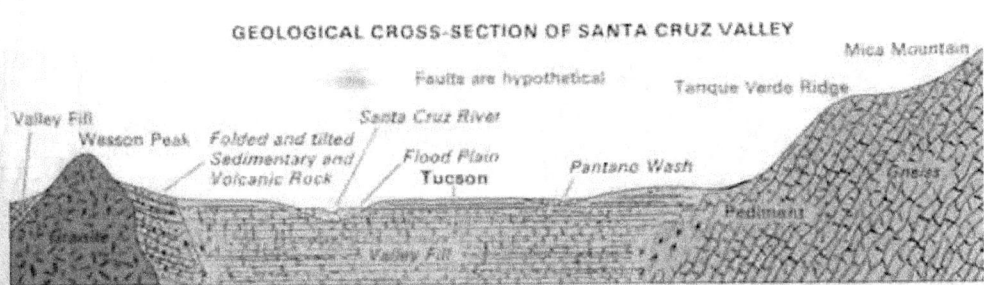

GEOLOGICAL CROSS-SECTION OF SANTA CRUZ VALLEY

The presence of the limestones and sandstones indicates that that area was submerged below the water of ancient seas one or more times in the geologic past. The gneiss, schist, and granite bespeak deep-seated metamorphism and magmatic intrusion, which gave these rocks the form, composition, and texture they possess today. The lava flows are indicative of volcanic activity that was a part of the extensive volcanism that occurred in this part of southern Arizona. And, finally, the great alluvial fans and bajadas suggest that these mountain ranges could bury themselves in their own products of erosion unless the mountain-building processes in future eons continue at a faster pace than the wearing-down processes. This evidence enables us to perceive today's landscape as but a transitional phase in the drama of change that will continue for milleniums.

During summer, mule deer and javelina frequent King Canyon, which has the only known permanent spring in the Tucson Mountains. The predominant rock in this major canyon northeast of the Arizona-Sonora Desert Museum is called Amole Arkose; it is of sedimentary origin and was deposited during the Cretaceous period.

17

Chiminea Creek, originating in the higher elevations of the Rincon Mountains, forms a ribbonlike oasis of hydrophytic plants, including large trees such as sycamore, walnut, and cottonwood.

plant and animal zones

Mountains are biologically exciting because their elevation produces, in a short vertical distance, the same climatic changes that occur over long latitudinal distances. In Arizona, average temperatures drop about 4° with each 1,000-foot rise in elevation, and precipitation increases about 4 to 5 inches. At sea level, you would have to travel about 900 miles north to experience the climatic change found in going from the lowest part of the monument (2,200 feet) to the top of Mica Mountain (8,600 feet). Plant-and-animal communities change along these climatic gradients.

In southern Arizona, the sequence of vegetation types begins with desert scrub at the lowest elevations, and ranges through desert grassland, oak woodland and chaparral, oak-pine woodland, ponderosa pine forest, and Douglas-fir forest, to Engelmann spruce forest on top of the highest mountains. In the national monument, we have all these types except the last—although desert grassland is poorly represented here because the steep slopes squeeze it into a narrow altitudinal band; and Douglas-fir forest is restricted to small areas in canyons and on north-facing slopes at high elevations. A few drainageways (notably Chiminea Canyon, on the south side of the monument) support patches of deciduous trees classified as riparian woodland.

These vegetation zones should not be visualized as nice neat bands on the mountainsides, however, for roughness of topography and differing tolerances of plant species usually lead to gradual changes from one zone to the next. Because of the greater exposure to drying heat and sunlight, each belt occupies a higher range of elevation on south-facing slopes than on north-facing ones. And along draws, where conditions are wetter and cooler than on ridges at the same elevations, the vegetation zones finger down to lower elevations; on ridges the opposite is true.

Animals are less restricted than plants to particular vegetation belts, but they, too, show a zonation with altitude. You must look, for instance, in the paloverde-saguaro community (part of the desert scrub type) for kangaroo rats and cactus wrens; in the oak-pine woodland for Mexican jays; and in the ponderosa pine forest for whitetail deer and tassel-eared squirrels.

19

In the Tucson Mountains, only the desert scrub type is well represented, although the highest ridges support a suggestion of desert grassland. A small patch of shrub live oaks, relicts of a wetter period, huddles on the north side of Wasson Peak. Zonation is poorly developed here because the mountains are low and small in mass. (Studies have shown that mountains of smaller mass tend to have warmer and drier climates than bulkier mountains of the same height.)

Major Vegetation Types in Saguaro National Monument

Vegetation type	Average July temp.	Elevations (feet)	Annual rainfall (inches)	Prominent species	Life zone
Southwestern Desert Scrub	94	2,200 to 4,000	7 to 13	Paloverde Saguaro Pricklypear Cholla Catclaw Ocotillo Creosotebush Mesquite Ironwood*	Lower Sonoran
Grassland Transition	85	3,500 to 4,500	10 to 15	Mesquite Beargrass Gramagrasses Amole Sotol	Upper Sonoran
Oak-pine-juniper Woodland and Chaparral	74	4,500 to 7,000	12 to 22	Emory Oak Mexican Blue Oak Shrub Live Oak Pinyon Pine Mountain-mahogany Manzanita Skunkbush	Upper Sonoran
Coniferous Forest	68	above 6,000	18 to 30	Ponderosa Pine Gambel Oak Buckbrush Mountain Muhly Douglas-fir White Fir Aspen Snowberry Mexican White Pine	Transition and start of Canadian

*TUCSON MOUNTAIN SECTION

Biologists have developed several systems for classifying assemblages of plants and animals over broad regions, and one of the most widely known is that of C. Hart Merriam. Around the turn of the century, he conceived a system of "life zones," named for the parts of the continent where they are best developed: Tropical, Lower Sonoran, Upper Sonoran, Transition, Canadian, Hudsonian, and Arctic. In Saguaro National Monument, the Lower Sonoran Zone corresponds to the desert scrub type; the Upper Sonoran includes desert grassland transition, oak woodland, and oak-pine woodland; the Transition Zone is equivalent to the ponderosa pine forest; and the poorly represented Canadian Zone has Douglas-fir and white fir. Because Merriam's system is so widely used, his terms are included in the tabulation on page 20, which summarizes the main characteristics of plant zones in the monument.

The best way to appreciate the biotic changes that occur with elevation is to walk or ride a horse to the top of the Rincons. But if your time or energy is limited you can get a quick view of these sequences by driving up nearby Mount Lemmon, in the Santa Catalina Mountains.

Saguaro Forest landscape from the scenic drive.

desert plants

When you look out over the cactus forest, in either part of Saguaro National Monument, you may think there's a sameness to it in all directions—saguaros standing amid scattered shrubs. But look harder, or walk about, and you will discover variations in the scene. First, there is a gradual change in the vegetation from the mountain foot down the bajada or pediment to the valley floor, as saguaros and paloverdes (green) become sparser and creosotebushes (smaller and brownish) take over. (The monument itself does not extend far enough from the mountain bases to include extensive creosotebush communities, but these cover large areas in the lowest parts of the valleys.) Then there is the luxuriant growth along washes, where mesquites and paloverdes grow to tree size and there are more kinds of plants. And if you are observant you may notice slight variations with each change in slope on the rolling hills—for example, more grass growing on their north sides. On another scale, you can see separate little communities of plants in special situations, as under shrubs or on rocks.

These patterns are due to variations in the environment. For the desert is hotter and drier in some places than in others. The gradual downslope changes in vegetation reflect the decrease in the amount of soil moisture available to plants—a condition caused by the decreasing size of soil particles and consequent shrinkage of water-holding space between particles. Desert washes encourage plant growth because they channel water and cold air. Strangely enough, night temperatures are often lower in a desert valley than farther up the slopes. This "inversion" is due to drainage of cold air down mountainsides, forming cold "pools" in valley bottoms, especially in winter. Cold air is heavier than warm, and (since air flows much like water) it is channeled down the draws— a fact that will strike you if you walk into a wash near the mountains at night or early in the morning. Add this phenomenon to the great amounts of moisture that lie beneath the surface of washes, and you can see that desert drainageways are really linear oases. More subtly, desert hills reflect in their vegetation the differences in soil moisture from north slopes to south slopes caused by increasing exposure to sunlight.

One controlling factor, then, is dominant in the desert: the

23

scarcity of water. This results not only in the unusual forms and adaptations of desert plants, but also in a distinctive type of plant community. In wetter climes, plants compete mostly for sunlight. They can grow close together, as long as they receive adequate illumination. During the regrowth of forests, several distinct sets of plants appear, each succeeding group more tolerant of shade than the last, as the forest canopy closes. In the desert, there is no such succession. Clear a patch of desert vegetation, and the same species will reappear—spaced out, with bare ground between them as before. For here sunlight is abundant (we might say over-abundant), but there is not enough water to allow plants to cover the ground.

Dr. Forrest Shreve, who was a botanist at the Carnegie Institution's Desert Laboratory near Tucson and a master student of deserts, defined a desert as "an area in which deficient and un-certain rainfall . . . has made a strong impression on the structure, functions, and behavior of living things." The distinctive charac-teristics of desert plants and animals have evolved through millions of years, in a trial-and-error process in which only the better-fitted forms have survived. It would be enlightening to know how many of the species and varieties of plants that developed during the past 60 million years or so have failed to adapt to Sonoran Desert conditions. It is fascinating to study the hundreds of forms that have succeeded and to try to determine what structures they have perfected and what methods they have originated that enabled them to maintain themselves in such a harsh environment.

Desert plants can be classified as "escapers," "evaders," and "resisters," according to their means of adaptation to water short-age. Escapers, such as the annuals, avoid the problem entirely by waiting out the dry periods as seeds, to sprout and reproduce only when the rains come. Evaders, such as the ocotillo, reduce their water loss during droughts by such methods as dropping their leaves or going into a state of dormancy. Resisters, however, "hang in there" all year, taking the desert's worst. The cactuses, prime examples of this group, rapidly soak up water from each rain and store it for use during drought; the mesquite's deep roots tap a more constant source of moisture.

Succulents

A large group of desert plants conserve water for use in periods of drought by storing it in specialized tissues during the wet season. Some of these "succulents," principally the yuccas, eschevarias, and agaves, have developed water-storage tissues in their leaves. A few, notably the NIGHTBLOOMING CEREUS (see appendix for scientific names of plants), have slender stems but an enormous, carrot-shaped root in which the moisture-storage tissue is located. The GOURDS also have large, thick, moisture-

25

retaining roots, as does the WILD-CUCUMBER.

The cactuses store water in their stems and thus are called stem succulents. In the course of its evolution the cactus has eliminated leaves, and their function has been taken over by the green outer covering of the stems. Thus the amount of transpiration (moisture loss through plant breathing pores) has been greatly reduced.

The cactuses are thought to have evolved from relatives of the rose family in the West Indies, beginning some 18,000 to 20,000 years ago. From there they spread to most parts of the Western

Saguaro buds.

Hemisphere, but particularly to the drier regions, changing their forms to meet new conditions. One of the youngest of plant families, the cactuses are still evolving rapidly. This doesn't make the task of classifying them easier for the taxonomists.

In varied forms, cactuses enliven the paloverde-saguaro community. In size they range from tiny button and pincushion types, some of which weigh only a few ounces, to the giant saguaro, the bulkiest of which may weigh several tons.

Cactuses as a group are easily recognized, although many people mistakenly believe that any desert plant with spines or thorns is a cactus. Shreve describes their main characteristics thus:

26

Several structural features have served to give the cacti their outstanding appearance, so unlike that of other plants. Most general have been the loss of the leaf as a permanent organ, the enlargement of the stem to accommodate water-storing tissue, and the development of local spinebearing structures known as "areoles." In several genera, the stem is segmented into sections which are flat and somewhat leaflike; in others the stem is round, much branched, and the surface occupied by closeset tubercles. In a large group, including massive erect forms, as well as slender climbing ones, the stem is grooved or fluted and thus able readily to accommodate its surface to great fluctuations in the water content of the tissues.

If you are trying to identify species, however, cactuses can be annoying, since they often hybridize. You must expect to find some individuals that don't fit the book descriptions.

The Saguaro—Monarch of the Monument

Of all the species of cactus recorded in Saguaro National Monument, the giant SAGUARO (pronounced sah-WAH-roe) holds the center of interest. From the visitor's standpoint, all other plants, no matter how bizarre in appearance or peculiar in living habits, are merely stage scenery for setting off the star of the desert drama. For size, this vegetable mammoth tops all other succulents of this country; heights of more than 50 feet and weights of more than 5 tons have been reported. There have been specimens with more than 50 arms, or branches. Although no accurate method of determining saguaro age has been devised, it is estimated that an occasional veteran may reach the two-century mark.

Structurally, the giant cactus is well adapted to meet the stern requirements of its habitat. Its widespread root system, as much as 70 feet in diameter, lying close to the surface of the ground, anchors and holds the heavy plant erect. The shallow root system quickly and efficiently collects and channels to the main stem any moisture that may penetrate the topsoil. The trunk and branches have a cylindrical framework of long slender poles or ribs fused at the constricted base. This woody skeleton supports the mass of pulpy tissue, the whole being covered with tough, waxy, spine-bearing "skin." Numerous vertical ridges, like the pleats of a huge accordion, permit the stems and branches to expand in girth as the tissues swell with water during wet weather and to shrink during times of drought.

So efficient is the saguaro's water-storage system that, even after years of extreme drought, the plant retains enough moisture in reserve to enable it to produce flower buds. The buds appear in vertical rows at the tips of the main stem and branches, a few opening each evening over a period of several weeks in May and June. The flowers, up to 4 inches in diameter, have waxy white

27

petals. This beautiful blossom is the State flower of Arizona. The egg-shaped fruits mature in late June and July, splitting open when ripe to reveal masses of juicy, deep-red pulp filled with tiny black seeds. Pulp and seeds are consumed by several kinds of birds, especially white-winged doves. Many fruits that fall to the ground are promptly eaten by rodents and other animals.

Indians, too, eat the fruits. European explorers who followed Coronado's expedition into this region found peaceful Papagos and Opatas living here, hunting animals and utilizing many native plants. Among the most dependable of Papago food sources was the fruit of the giant cactus. So important was this fruit harvest in their economy that they designated this season as the start of the new year. Today, in some parts of the desert, Pima and Papago Indians still harvest the fruits. The word "saguaro" is believed to derive from a Spanish corruption of a Papago word for the big cactus.

Saguaros provide not only food for man and beast, but homes for animals. Walk through a giant cactus forest and you will be

Spiny armor of the saguaro (left).   Saguaro in full bloom.

amazed at the number of holes drilled in these plants. The holes are made by Gila woodpeckers and related gilded flickers, which often relinquish them after one nesting season. The next occupant may be any of a host of desert dwellers, including screech and elf owls, purple martins, and invertebrates. Some birds use the plant as a foundation for their homes. White-winged doves, for instance, often build flimsy stick platforms on the tips of saguaro arms; red-tailed hawks and horned owls construct more substantial nests in the forks.

Although billions of saguaro seeds are produced yearly in the extensive stands of the monument, only a very few find favorable locations for germination and growth. Trees, rocks, dead limbs, pebbles—anything that reduces evaporation in the immediate vicinity of the seed—improve the chance of germination. These kinds of shelter not only provide the necessary moisture conditions, but also hide the seed from armies of ants, rodents, and other animals searching for food.

Early growth is extremely slow. A 2-year-old saguaro may be

Saguaro blossoms (left). Saguaro fruit.

Gila woodpecker at its nesting hole.

only one-quarter of an inch in diameter, and a 9-year-old plant may be 6 inches high. These years are the most hazardous. Insect larvae devour the tiny cactuses. Woodrats and other rodents chew the succulent tissue for its water, and ground squirrels uproot the young plants with their digging. In later life, the saguaro must contend with uprooting wind and human vandalism, as well as the earlier foes—drought, frost, erosion, and animals.

In a century of maturity, a saguaro may produce 50 million seeds; replacement of the parent plant would require only that one of these germinate and grow. But in the cactus forest of the Rincon Mountain Section, the rate of survival has been even lower, so that over the last few decades the stand has been dwindling. What is wrong?

Many answers to this question have been advanced, but like all interrelationships in nature, the saguaro's role in the desert web of life is very complex, and involves past events as well as present ones; a partial answer to the problem may be all we can hope for. The following reasons for the decline of the saguaros have been suggested by researchers.

Saguaro, 1 foot high, in a rocky habitat.   A typical 4-foot saguaro.

There is some evidence to suggest that the Southwest has been getting drier since at least the late 19th century, and while the saguaro is adapted to extreme aridity, some of the "nurse" plants that shelter it during infancy are not. If such shrubs as paloverdes and mesquites dwindle, it is argued, so must the saguaro, which in its early years depends on them for shade.

Other culprits in the saguaro problem are man himself and his livestock. Around 1880, soon after the first railroad reached Tucson, a cattle boom began in southern Arizona. The valleys were soon overstocked, and cattle scoured the mountainsides in search of food. By 1893, when drought and starvation decimated the herds, the land had been severely overgrazed. Though the monument was established in 1933, grazing in the Rincon Mountain Section's main cactus forest continued until 1958. (Elsewhere in the monument, it still goes on.) Compounding the problem, woodcutters removed acres of mesquite and other trees. In the center of the present Cactus Forest Loop Drive, lime kilns devoured quantities of woody fuel. Further upsetting the desert's natural balance, ranchers and Government agents poisoned coyotes

31

and shot hawks and other predators—in the belief that this would benefit the owners of livestock.

This unrestrained assault on the environment had unfortunate effects on saguaros as well as on the human economy. Overgrazing may have resulted in an increase in kangaroo rats (which benefit from bare ground on which to hunt seeds) and certain other rodents adapted to an open sort of ground cover. Man's killing of predators, their natural enemies, further encouraged proliferation of these rodents, which some people say are especially destructive of saguaro seeds and young plants. Whatever the effect these rodents have on the saguaros, the removal of ground cover intensified erosion and reduced the chances for the seeds to germinate and grow. And certainly the cutting of desert trees removed shade that would have benefited young saguaros. In the Tucson Mountain Section, which is near the northeastern edge of the Sonoran Desert, freezing temperatures are perhaps the most important environmental factor in saguaro mortality.

Although the causes of decline of the cactus forest lying northwest of Tanque Verde Ridge are still something of a puzzle, several facts are clear: the saguaro is not becoming extinct; in rocky habitats many young saguaros are surviving, promising continued stands for the future; in non-rocky habitats, some young saguaros are surviving, ensuring that at least thin stands will endure in

Looking toward the Santa Catalina Mountains from Cactus Forest Drive in September 1942.

32

these areas. Furthermore, since grazing was stopped here, ground cover has improved—a plus factor for the saguaro's welfare. On the negative side, it is possible that, in addition to suffering from climatic, biotic, and human pressures, the once-dense mature stands of the monument are in the down-phase of a natural fluctuation. It is possible, too, that these stands owed their exceptional richness to an unusually favorable past environment which may not occur again. We can hope, however, that sometime in the not-distant future the total environmental balance will shift once again in favor of the giant cactus.

Other Common Cactuses

Many other cactuses share the saguaro's environment. The BARREL CACTUS is sometimes mistaken for a young saguaro, but can easily be distinguished by its curved red spines. Stocky and unbranching, this cactus rarely attains a height of more than 5 or 6 feet. It bears clusters of sharp spines, called "areoles," with the stout central spine flattened and curved like a fishhook. In bloom, in late summer or early autumn, this succulent plant produces clusters of yellow or orange flowers on its crown. The widely circulated story that water can be obtained by tapping the barrel cactus has little basis in fact, although it is possible that the thick, bitter juice squeezed from the plant's moist tissues might, under extreme conditions, prevent death from thirst. Desert rats,

A photograph taken from the same spot in January 1970.

mice, and rabbits, carefully avoiding the spines, sometimes gnaw into the plant's tissues to obtain moisture.

The group of cactuses called opuntias (oh-POON-cha) have jointed stems and branches. They are common and widespread throughout the desert and are well represented in the monument.

Those having cylindrical joints are known as chollas (CHO-yah), while those with flat or padlike joints are called pricklypears.

Chollas range in size and form from low mats to small trees, but most of those in the monument are shrublike. TEDDY BEAR CHOLLA, infamous for its barbed, hard-to-remove-from-your-skin spines, forms thick stands on warm south- or west-facing slopes. Its dense armor of straw-colored spines and its black trunk identify it. Because its joints break off easily when in contact with man or animal, this uncuddly customer is popularly called "jumping cactus." A similar species is CHAIN FRUIT CHOLLA, notable for its long, branched chains of fruit, which sometimes extend to the ground. Each year, the new flowers blossom from the persistent fruit of the previous year. There is a common variety of this species that is almost spineless. STAGHORN CHOLLA, an inhabitant primarily of washes and other damp places, is named from its antler-shaped stems. This cactus' scientific name—*Opuntia versicolor*—refers to the fact that its flowers, which appear in April and May, may be yellow, red, green, or brown. (Each plant sticks with one color through its lifetime.) Among the smaller chollas, thin-stemmed PENCIL CHOLLA grows from 2 to 4 feet high on plains and sandy washes. DESERT CHRISTMAS CACTUS, almost matlike in form, blooms in late spring but develops brilliant red fruits which last through the winter.

PRICKLYPEARS, like many of the chollas, produce large blossoms

*Barrel cactus blossoms (left). Barrel cactus spines (center).
Chain fruit cholla (right) at Tucson Mountain Section headquarters.*

in late spring. Those on the monument are principally the yellow-flowered species. The reddish brown-to-mahogany colored edible fruits, called tunas, attain the size of large strawberries. When mature in autumn, they are consumed by many desert animals.

Some of the smaller cactuses are so tiny as to be unnoticeable except when in bloom; examples are the HEDGEHOGS, the FISH-HOOKS, and the PINCUSHIONS. Blossoms of some of these ground-hugging species are large, in some cases larger than the rest of the plant, and spectacular in form and color. All add to the monument's spring and early summer display of floral beauty.

Non-Succulents

For the diversity of devices for adaptation to an inhospitable environment, the many species making up the non-succulent desert vegetation provide an absorbing field for study. As we have seen, there are two ways to survive the harsh desert climate; one is to avoid the periods of excessive heat and drought ("escapers"); the other is to adopt various protective devices ("evaders" and "resisters"). Short-lived plants follow the first method; perennials, the second.

Perennials

Chief among the requirements for year-round survival in the desert is a plant's ability to control transpiration and thus maintain a balance between water loss and water supply. In this struggle, the hours of darkness are a great aid because in the cool of the night the air is unable to take up as much moisture as it does under the influence of the sun's evaporating heat. Therefore, less exhaling and evaporating of water occurs from plants, and both the rate and the amount of water loss are reduced. This reduction in transpiration at night allows the plants to recover from the severe drying effects of the day. One biologist may have been close to the truth when he stated, "If the celestial machinery should break down so that just one night were omitted in the midst of a dry season, it would spell the doom of half the nonsucculent plants in the desert."

One of the common trees in the desert part of the monument is the MESQUITE (mess-KEET). In general appearance it resembles a small, spiny apple or peach tree with finely divided leaves. Its roots sometimes penetrate to a depth of 40 or more feet, thus securing moisture at the deeper, cooler soil levels, from a supply that remains nearly constant throughout the year. This enables the tree to expose a rather large expanse of leaf surface without losing more water than it can replace. A number of mechanical devices help the tree reduce its water loss during the driest part of the day (10 a.m. to 4 p.m.). Among these are its ability to fold its leaves and close the stomata (breathing pores), thereby greatly reducing the surface area exposed to exhaling and evaporating influences. In April and May, mesquite trees are covered with pale-yellow, catkinlike flowers which attract swarms of insects. These flowers develop to stringbeanlike pods rich in sugar and important as food for deer and other animals. In earlier days, the mesquite was also a valuable source of food and firewood for Indians and pioneers.

Another desert tree abundant in the monument is the YELLOW PALOVERDE. It is somewhat similar in size and general shape to the mesquite. Lacking the deeply penetrating root system of the mesquite, the paloverde (Spanish word meaning "green stick") has no dependable moisture source; but it has made unusual adaptations that enable it to retain as much as possible of the water collected by its roots. In early spring the tree leafs out in dense foliage, which is followed closely by a blanket of yellow blossoms. At this season the paloverdes provide one of the most spectacular displays of the desert, particularly along washes, where

Top row: Pricklypear blossom (left). Claret cup hedgehog (center). Fishhook cactus. Bottom row: Cholla in bloom (left). Staghorn cholla.

they grow especially well. Blue paloverde, growing in the arroyos, blooms well every year. Yellow, or foothill, paloverde, a separate species, blooms only if the soil moisture is high following winter rains.

With the coming of the hot, drying weather of late spring, the trees need to reduce their moisture losses. They gradually drop their leaves until, by early summer, each tree has become practically bare. The trees do not enter a period of dormancy, but are able to remain active because their green bark contains chlorophyll. Thus, the bark takes over some of the food-manufacturing function normally performed by leaves, but without the high rate of water loss.

Carrying the drought-evasion habits of the paloverde a step further, the OCOTILLO (oh-koh-TEE-yoh) comes into full leaf following each rainy spell during the warmer months. During the intervening dry periods it sheds its foliage. The ocotillo, a common and conspicuous desert dweller, is a shrub of striking appearance, with thorny, whiplike, unbranching stems 8 to 12 feet long growing upward in a funnel-shaped cluster. In spring, showy scarlet flower clusters appear at the tips of the stems, making each plant a glowing splash of color.

A number of desert shrubs fail to display as much ingenuity as the paloverde. Some of these evade the dry season simply by going into a state of dormancy. The WOLFBERRY bursts into full leaf soon after the first winter rains and blossoms as early as January. Its small, tomato-red, juicy fruits are sought by birds, which also find protective cover for their nests and for overnight perches in the stiff, thorny shrubs. In the past, the berrylike fruits

Mesquite in bloom.

were important to the Indians, who ate them raw or made them into a sauce.

Commonest of the conspicuous desert non-succulent shrubs is the wispy-looking but tough CREOSOTEBUSH, found principally on poor soils and on the desert flats between mountain ranges. It is also sprinkled throughout the paloverde-saguaro community in the monument. A new crop of wax-coated, musty-smelling leaves, giving the plant the local (but mistaken) name "greasewood," appears as early as January. The leaves are followed by a profuse blooming of small yellow flowers and cottony seed balls. During abnormally moist summers or in damp locations, the leaves and flowers persist the year round; but usually the coming of dry

Yellow paloverde, Tucson Mountain Section.

Ironwood blossoms (left).   *Parry's penstemon.*

weather brings an end to the blossoming period. If the dry spell
is exceptionally long, the leaves turn brown, and the plants remain
dormant until awakened by the next winter's rainfall. Pima Indians
formerly gathered a resinous material, known as lac, which accum-
ulates on the bark of its branches, and used it to mend pottery and
fasten arrow points. They also steeped the leaves to obtain an
antiseptic medicine. Ground squirrels commonly feed on the seeds.

A large shrub of open, sprawling growth usually found along
desert washes in company with mesquite is CATCLAW. Its name
refers to the small curved thorns that hide on its branches. In
April and May, the small trees are covered with fragrant, pale-
yellow, catkinlike flower clusters that attract swarms of insects.
The seed pods were ground into meal by the Indians and eaten
as mush and cakes.

In lower elevations of the Tucson Mountain Section, the gray-
blue foliage of IRONWOOD is a common sight, but the species does
not range farther eastward. Its wisterialike lavender-and-white
flowers blossom in May and June. The nutritious seeds are harvested by rodents and formerly were parched and eaten by Indians. The wood is so dense that it sinks in water; Indians used
it for making arrowheads and tool handles.

Ferns—commonly, plants of dank woods and other moist habitats—seem entirely out of place in the desert; nevertheless, some
members of the fern family have overcome drought conditions.
The GOLDFERN is common on rocky ledges, where it persists by
means of special drought-resistant cells.

Among the smaller perennials are many that add to spring
flower displays when conditions of moisture and temperature are

40

favorable. Perennials do not need to mature their seeds before the coming of summer as do the ephemerals; a majority start blossoming somewhat later in the spring, and gaily flaunt their flowers long after the annuals have faded and died. When the heat and drought of early summer begin to bear down, they gradually die back, surviving the "long dry" by their persistent roots and larger stems. One of the most noticeable and beautiful of this group of small perennials fairly common in the monument is PARRY'S PENSTEMON. It occurs in scattered clumps on well-drained slopes along the base of Tanque Verde Ridge. The showy rose-magenta flowers and the glossy-green leaves arise from erect stems that may grow 4 feet tall in favorable seasons.

Among the first of the shrubby perennials to cover the rocky hillsides with a blanket of winter and springtime bloom is the BRITTLEBUSH. Masses of yellow sunflowerlike blossoms are borne on long stems that exude a gum which was chewed by the Indians and was also burned as incense in early mission churches.

A conspicuous perennial that survives the dry season as an underground bulb is BLUEDICKS. Although it doesn't occur in massed bloom, it does add spots of color to the desert scene. Usually appearing from February to May, bluedicks has violet flower clusters on long, slender, erect stems. The bulbs were dug and eaten by Pima and Papago Indians.

Although neither conspicuous nor attractive, the common TRI-ANGLE BURSAGE is an important part of the paloverde-saguaro community in the Tucson Mountains. A low, rounded, white-barked shrub, bursage has small, colorless flowers without petals. (Being wind-pollinated, the flowers do not need to attract insects.)

One of the handsome shrubs abundant in the high desert along the base of Tanque Verde Ridge is the JOJOBA (ho-HOH-bah), or deernut. Its thick, leathery, evergreen leaves are especially noticeable in winter and furnish excellent browse for deer. The flowers are small and yellowish, but the nutlike fruits are large and edible, although bitter. They were eaten raw or parched by the Indians, and were pulverized by early-day settlers for use as a coffee substitute.

Among the attractive flowering shrubs are the INDIGOBUSHES, of which there are several species adapted to the desert environment. The local, low-growing indigobushes are especially ornamental when covered with masses of deep-blue flowers in spring.

Another small shrub, noticeable from February to May because of its large, tassel-like pink-to-red blossoms and its fernlike leaves is FAIRY-DUSTER. Deer browse on its delicate foliage.

The PAPER FLOWER, growing in dome-shaped clumps covered with yellow flowers, sometimes blooms throughout the entire year. The petals bleach and dry and may remain on the plant weeks after the blossoms have faded.

Quick to attract attention because of their apparent lack of foliage, the JOINTFIRS, of which there are several desert species, grow in clumps of harsh, stringy, yellow-green, erect stems. The skin or outer bark of the stems performs the usual functions of leaves, which on these plants have been reduced to scales. Small, fragrant, yellow blossom clusters, appearing at the stem joints in spring, are visited by insects attracted to their nectar.

Ephemerals

Every spring, after a winter of normal rainfall, parts of the southwestern deserts are carpeted with a lush blanket of fast-growing annual herbs and wildflowers—the early spring ephemerals. The monument does not get massive displays, however, since it is lacking in the species that make the best show. But it does have many annuals that are beautiful individually or in small groups. Many of these "quickies" do not have the characteristics of desert plants; some of them, in fact, are part of the common vegetation of other climes where moisture is plentiful and summer temperatures are much less severe.

What are these "foreign" plants doing in the desert, and how do they survive? With its often frostfree winter climate and its normal December-to-March rains, the desert presents in early spring ideal growing weather for annuals that are able to compress a generation into several months. Several hundred species of plants have taken advantage of this situation.

There is WILD CARROT, which is a summer plant in South Carolina and a winter annual in California (where it is called "rattlesnake weed"). In the desert, its seeds lie dormant in the soil through the long, hot summer and the drying weather of autumn. Then, under the influence of winter rains and the soil-warming effects of early spring sunshine, they burst into rapid growth. One of a host of species, this early spring ephemeral is enabled by these favorable conditions to flower and mature its seed before the pall of summer heat and drought descends upon the desert. With their task complete, the parents wither and die. Their ripened seeds are scattered over the desert until winter rains enable them to cover the desert with another multicolored but short-lived carpet of foliage and bloom.

The one-season ephemerals do not limit themselves to the winter growing period. From July to September, local thundershowers deluge parts of the desert while other areas, not so fortunate, remain dry. Where rain has fallen, another and entirely different group of plants, called the summer ephemerals, find ideal conditions for growth and take their turn at weaving a desert carpet. Their seeds have lain dormant over winter. These summer "quickies" are plants that, farther south in Baja California and Sonora, Mexico, flourish during the winter rainy season. Saguaro National Monument is

42

doubly fortunate in that it lies within a section of desert having not only its own year-round vegetation, but also summer wildflowers "borrowed," for winter use, from its eastern and western neighbors, and winter wildflowers for summer decorations from its southern neighbors.

The short-lived leafy plants of summer and winter are able to compress their entire active life into 6 to 12 weeks when conditions are most suitable. Thus, they can escape all the rigorous periods of the desert climate by living for 8 or 9 months in the dormant seed stage. Some of the spectacular and colorful flowers of the monument are among these ephemerals that survive desert

Phacelia, an ephemeral.

White tackstem.

conditions by escaping them. It must be remembered, however, that when drought conditions or abnormally cold spring weather upset the norm—a not unusual occurrence—response of ephemeral plants is greatly restricted. If suitable conditions do not develop during the season for growth of a particular kind of ephemeral, its seeds will simply wait a year or more until conditions are favorable.

How do the seeds of ephemerals "know" when it is time to germinate? Experiments have revealed that the seeds of annuals will germinate only when enough water percolates through the soil to dissolve away a "growth inhibitor" in their coats. A single light rain will not do the job. In this way premature sprouting into a too-dry environment is prevented. The winter annuals, furthermore, will only germinate when soils are cool, and the summer annuals when soils are warm. These finely tuned adaptations thus allow plantlife to take full advantage of favorable seasons in the desert.

Early spring ephemerals climax their show in March. From late February to mid-April they are completing their growth and putting forth the precious seeds that will assure survival for the next generation. At the head of this parade of flowers in the monument is a purple-blossomed immigrant from the Mediterranean, the now thoroughly naturalized FILAREE. In addition to the small purple flowers, which may appear as early as January, the conspicuous "tailed" fruits almost always attract attention. When dry, they are tightly twisted, corkscrewlike; when damp they uncoil,

44

Desert-marigold.

forcing the needle-tipped seeds into the soil.

INDIAN WHEAT is one of the first plants to lay a green carpet over the sandy desert floor in spring. The tan, individual flower heads are conspicuous, but their numerous, close-growing spikes form a thick, luxuriant, pilelike ground cover. The countless tiny seeds are eagerly sought each spring by coveys of Gambel's quail, and are also harvested by Pima and Papago Indians.

DESERT CHICORY is somewhat like the common yellow dandelion but is longer stemmed and less coarse. Its white or butter-yellow blossoms make it one of the noticeable spring annuals in the desert. It rarely grows in pure stands but appears in conspicuous clumps among other short-lived plants.

Somewhat similar in appearance to desert chicory is WHITE TACKSTEM, one of the handsomest of the spring quickies. It is usually found on dry, rocky hillsides and has white or rose-colored flowers. Its name is derived from the presence of small glands which protrude in the manner of tiny tacks partially driven into the stems.

Following abnormally wet winters, FIDDLENECK covers patches of sandy or gravelly soil with a dense growth of bristly erect plants. These bear tight clusters of small yellow-orange blossoms arranged along a curling flower stem resembling the scroll end of a violin, hence the name. This plant, favored by the same growing conditions as creosotebush, frequently forms a dense though short-lived growth around the bases of those shrubs.

Associated with fiddleneck and creosotebush, SCORPIONWEED

45

adds its violet-purple blooms to the spring flower display following winters of above-normal precipitation. The name is derived from the curling habit of the blossom heads, which may remind the observer of the flexed tail of a scorpion. Touching the plant may cause skin irritation in susceptible individuals. Unfortunately, scorpionweed is also widely known as wildheliotrope, thus contributing to the confusion engendered by duplication of popular names. The plant properly called WILD-HELIOTROPE is similar in general appearance, but the flowers are white to pale purple and their odor is more pleasing than that of the scorpionweed. Wildheliotrope, or "quailplant," is another of the early spring ephemerals, but under favorable conditions, where soils are moist, it may continue to live and bloom throughout the year.

Growing in sandy locations and quite noticeable because of its large, yellow, showy, long-stemmed flower heads, the DESERT-MARIGOLD helps to open the spring blossoming season. Where moisture conditions are favorable, these plants may continue to bloom throughout the summer and well into autumn. Sometimes, during the hottest, driest time of the year, desert-marigolds are among the very few blossoms brightening the desert floor. Their

Bladder-pod.

bleached, papery petals persist for days after the flowers have faded, giving the plant the name paper-daisy.

One of the few species that makes a carpet of color in the monument is the tiny BLADDER-POD. This low-growing annual of the mustard family begins to cover open stretches of desert with a yellow blanket in late February or early March following wet winters. Bladder-pod is usually found in pure stands surrounding islands of cholla, creosotebush, and paloverde. It also mingles with other spring ephemerals, where it is promptly submerged by the ranker, taller-growing, more conspicuous annuals.

Illustrating one of the interesting phases in evolutionary variations among plants, the LUPINES are represented by several species which are able to survive and prosper in the desert. Some of the lupines are annuals of the quickie type; others are perennials with a life cycle of several years. Some of these longer-lived species join the ephemerals in the spring flower show, while others are more leisurely in approaching their blossoming time.

The Desert Scrub Community *occupies the lower parts of the monument, from about 2,200 to 4,000 feet.*

The Grassland Transition *begins at about 3,500 feet (overlapping in altitude with the Desert Scrub) and extends to 4,500 feet.*

major plant communities
of saguaro

In any region where a great range of altitude exists, the vegetation grows in a continuum of overlapping but recognizable zones. Climate being the major controlling factor of this zonal distribution, the plants of each band grow higher on south-facing slopes than on. cooler, moister north slopes. In Saguaro the major plant communities roughly correspond to altitudinal bands—with the desert at the lower, drier levels, and the transition through grasses and

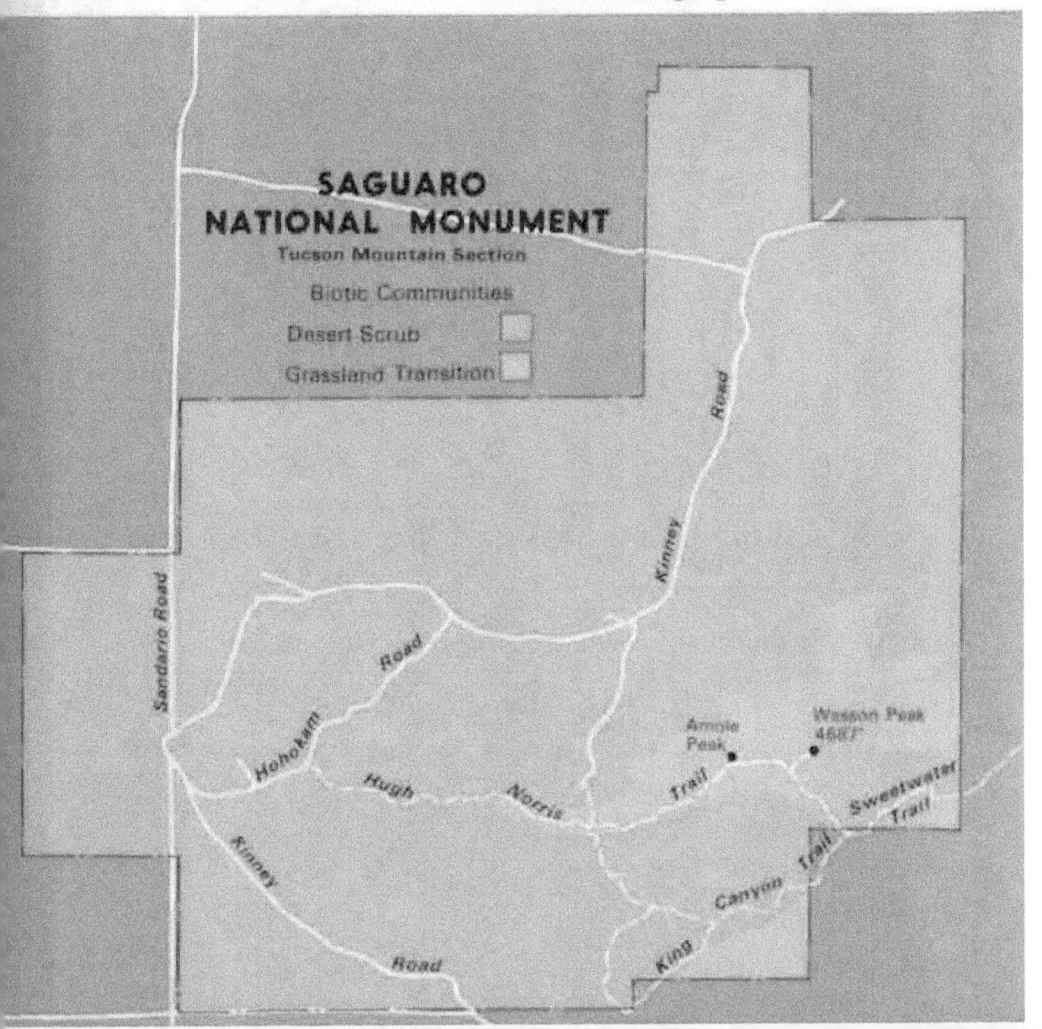

shrubland to forest community occurring with increasing altitude (and increasing rainfall). Timberline—the boundary between high mountain forests and the alpine meadows and barrens—does not exist in Saguaro, where trees clothe the highest peaks.

The fauna of each community is made up of animals adapted

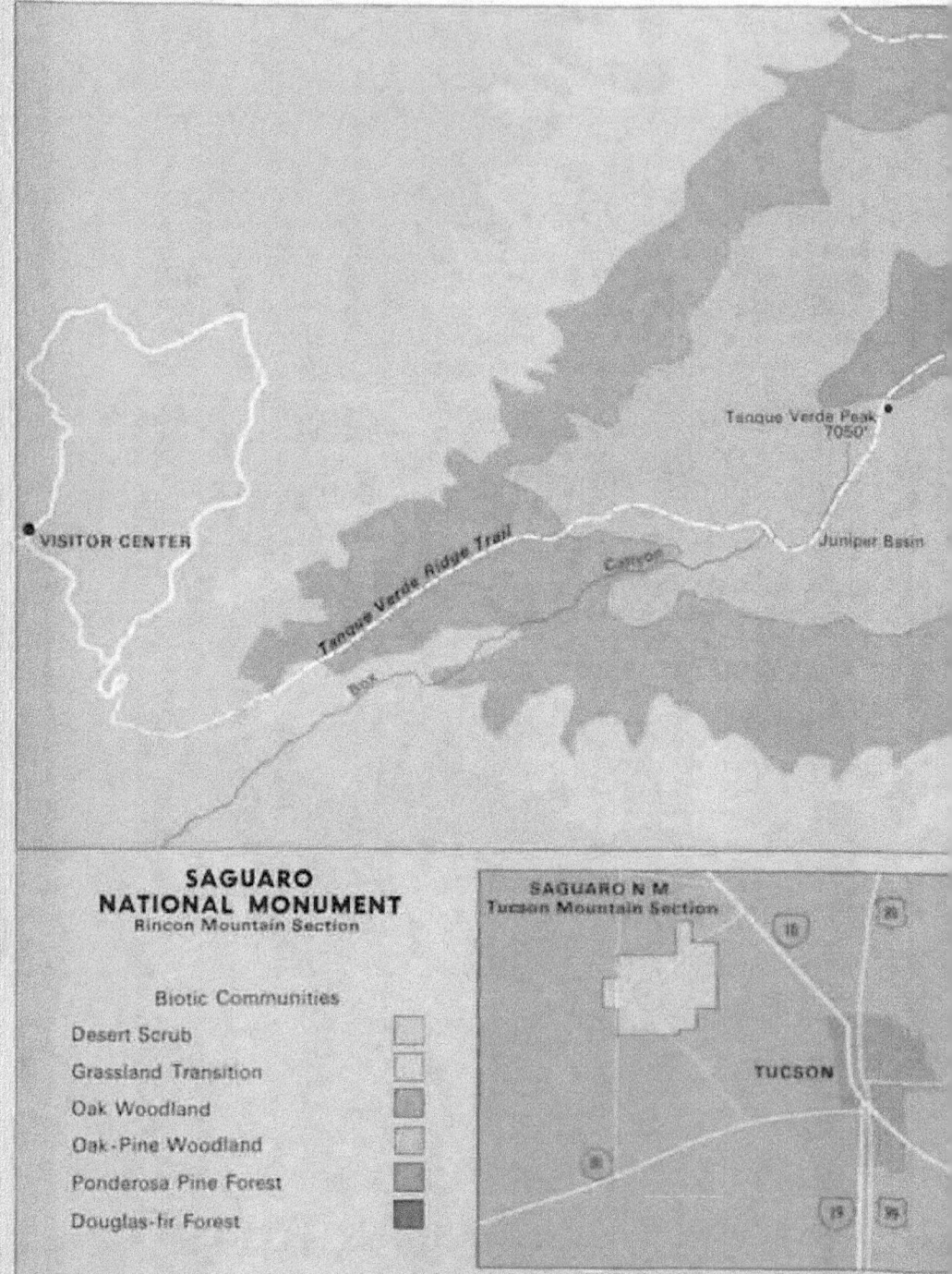

SAGUARO
NATIONAL MONUMENT
Rincon Mountain Section

Biotic Communities

Desert Scrub

Grassland Transition

Oak Woodland

Oak-Pine Woodland

Ponderosa Pine Forest

Douglas-fir Forest

to the climatic conditions and the available food, cover, and water. In Saguaro, some animals, such as the desert kangaroo rat and the cactus wren, are restricted by their life requirements to a narrow belt; others, such as the gray fox and great horned owl, are more adaptable and live in all the major plant communities.

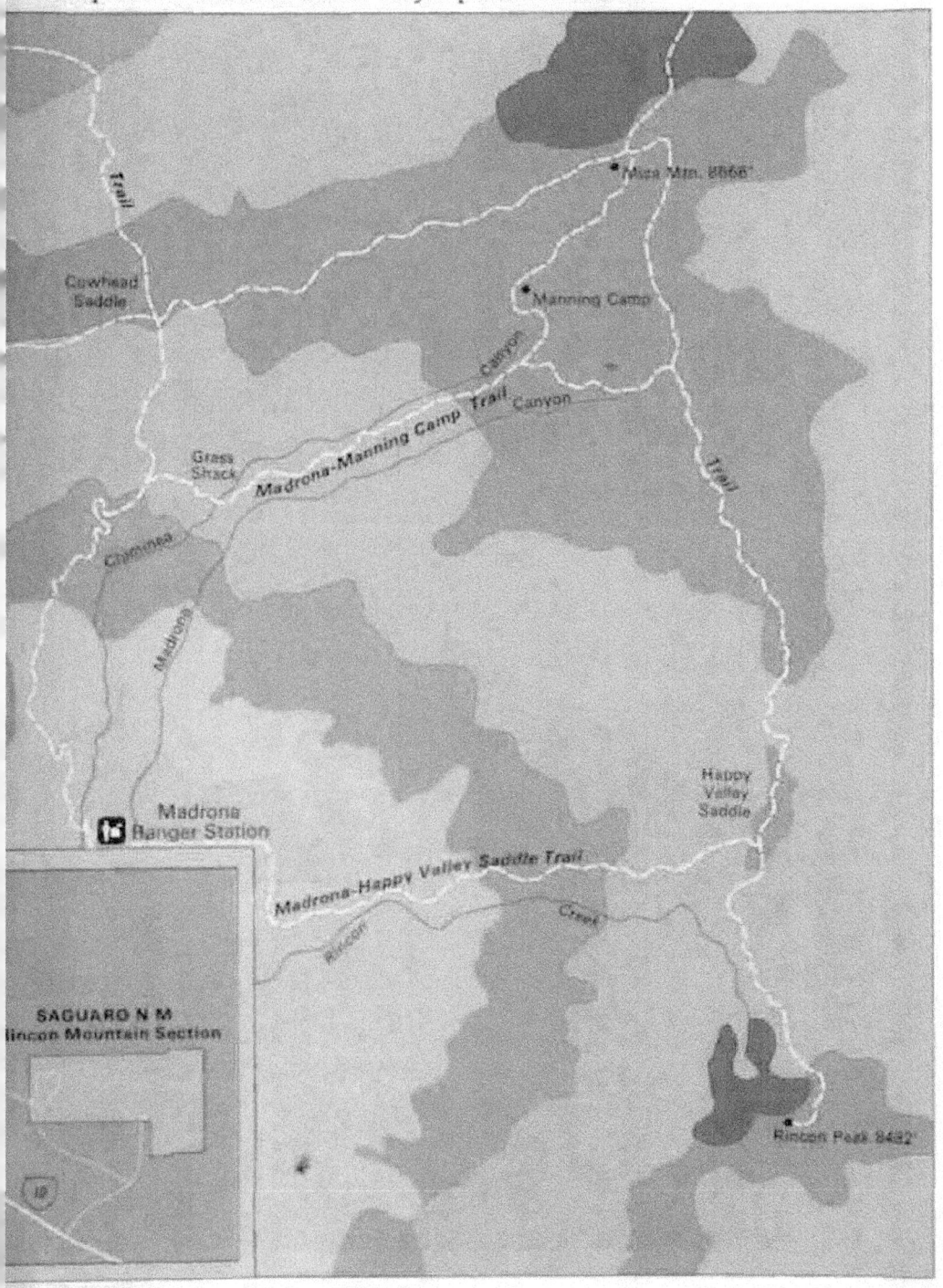

The Open Oak Woodland *has scattered trees associated with many plants of the Grassland Transition.*

The Oak-Pine Woodland, *from 4,500 to 7,000 feet, occupying a slightly cooler, wetter environment than those below, is broken by open glades and grassy hillsides.*

The Ponderosa Pine Forest, *unlike the Oak-Pine "pygmy woodland," has tall trees growing in clear, open stands. It covers much of the Rincon Mountains above 6,000 feet.*

The Douglas-fir Forest, *limited to the higher northern and northeastern slopes of Mica Mountain and the north side of Rincon Peak, often contains aspens and white firs.*

Spanish dagger, with Rincon Peak in the background.

plants of the hills and mountains

Most visitors to the monument see only a small part of it—the lowland cactus forests. But above the desert in the high back country of the Rincon Mountains is another world. These highlands are accessible only on foot or horseback; and they should remain so, for they are now in the last major roadless range in southern Arizona, and as such their wilderness value appreciates yearly. Interesting scenic trails reach the high places of the Tucson Mountains, too, but to see the greatest changes in plantlife you must climb the higher Rincons.

Oak-Pine Woodland

The thinning and final disappearance of saguaros along the trailside, although mesquite and ocotillos seem almost as numerous as on the floor of the desert, indicate that you are entering a slightly cooler, wetter environment. An occasional MEXICAN BLUE OAK or ARIZONA ROSEWOOD appears among the mesquites along the washes. The grasses that increasingly cover the ground as you climb include HAIRY, SIDEOATS, and SPRUCETOP GRAMA; CURLY MESQUITEGRASS; TANGLEHEAD; TEXAS BLUESTEM; and WOLFTAIL. Mingling with them are TURPENTINE-BUSH and shrubby SNAKEWEED.

A spectacular inhabitant of the grassland transition and open oak woodlands you are now entering is the AMOLE, or SHIN-DAGGER, whose rapidly growing blossom stalks attract attention from May to as late as August. The plants themselves, which grow crowded together in patches, consist of rosettes of succulent leaves superficially resembling bunches of flattened, green bananas. The stiff, needle-sharp leaf tips can inflict painful jabs on man and beast. During its lifetime, the plant stores food in its short, thick stem. Finally, after several years, it sends up an unbranched flower stalk that grows to 5 feet. The light-yellow flowers mature to brown, capsule-like fruits, after which the plant dies. The short stems or crowns, containing saponin, were used by Indians as soap. They also roasted the young bud stalks of some species by covering them with heated stones in pits.

Two noticeable plants of the lily family which sometimes dominate gravelly slopes of the grassland-oak woodland belt are the SOTOL and BEARGRASS, or sacahuista. The former grows from a

55

Mexican blue oak foliage.

compact crown as a dense, rather symmetrical cluster of long, thin ribbonlike leaves, usually grayed at the tips and armed along the margins with curved thorns. In early summer many small, cream-colored blossoms develop along the upper extremity of single fast-growing flower stalks 8 to 10 feet high. The bud stalks formerly were harvested and roasted by Indians. In Mexico a potent alcoholic drink, sotol, is distilled from the fermented juice of the pounded crowns. Sacahuista resembles huge, sprawling clusters of coarse grass. Flower stalks are short, producing conspicuous, open, loose sprays of small, tan-to-brownish flowers in May and June. Indians used the tender bud stalks for food and obtained fiber from the long, slender leaves, weaving them into baskets and mats.

Uncommon in the monument, but worthy of mention, are ARIZONA SYCAMORE and ARIZONA CYPRESS. The latter is restricted to the east flank of the Rincon Mountains, steep slopes, and deep canyons, where it grows with the SILVERLEAF, PALMER, EMORY, and NETLEAF OAKS; mesic shrubs; poison oak; and CALIFORNIA BUCKTHORN. Arizona sycamore grows along lower-canyon water-courses such as Chiminea and Rincon Creeks, which drain the rugged south flank of Mica Mountain and the west flank of Rincon Peak, respectively.

As you continue to climb, the open, grassy, shrub-dotted slopes change in places to sprawling thickets, called chaparral. These are made up of manzanita and skunkbush, SILKTASSEL, evergreen oaks, and underbrush of smaller shrubs. Among the common oaks are ARIZONA WHITE OAK and, on drier sites of the Tanque Verde range, SHRUB LIVE OAK. The oaks furnish protective cover, browse, and acorns for deer and other mammals and birds, and are of great

56

Turpentine-bush, a member of the sunflower family.

value in retarding soil erosion on steep gravelly slopes.

POINTLEAF MANZANITA is especially abundant on the lower eastern flanks of the Rincon Mountains in the Happy Valley area. Early in spring the waxy, urn-shaped blossoms, the leathery, glossy, evergreen leaves, and the typical grotesquely crooked, red-barked limbs, make manzanita one of the most attractive shrubs of the chaparral.

Although SKUNKBUSH is a close relative of poison ivy and sumac, its aromatic foilage is harmless. Growing in compact thickets, skunkbush provides food and cover for birds and other small animals. Inconspicuous yellow flowers appear from March to June, and are followed by berrylike fruits which are dull red when mature.

As you follow the trail higher, occasional MEXICAN PINYON PINES and ALLIGATOR JUNIPERS appear. Gradually these ever-greens become more abundant, mingling with the oaks to form a pigmy oak-pine-juniper forest. Clumps of MOUNTAIN-MAHOGANY are noticeable, their feathery seed "tails" gleaming in the sunshine.

Pinyons are among the commonest and most widespread trees of the middle elevations throughout the Southwest. The Mexican pinyon, which is the species growing abundantly in the Tanque Verde-Rincon upland, may be recognized by the fact that its foliage is in clusters of three needles to the bundle. Its cones require nearly 2 years to mature and contain hard-shelled seeds or nuts which are a source of food for many birds and mammals. These pines are usually shrubby, rarely more than 15 to 25 feet high, with horizontal, twisted, low-growing limbs. Intermingled with the pinyons are alligator junipers, often mistakenly called cedars. Those in the monument are conspicuous because their

57

platy bark forms an attractive pattern resembling the squarish-scaled skin of alligators. The berrylike cones are soft and mealy, and are eaten by many kinds of wild animals.

Although the oak-pine woodland supports a heavy stand of shrubby trees over much of the terrain, there are numerous open glades and grassy hillsides. In addition to some of the aforementioned grasses, BLUEGRASS, BULLGRASS, and PLAINS LIVEGRASS provide ground cover in this belt. Following summer showers, many flowering herbs brighten the open slopes. Yellow to orange petals of PUCCOON, and the white to lavender-and-rose blossoms of MOCK-PENNYROYAL and HOUSTONIA are among those seen along the trailside.

Ponderosa Pine Forest

Just as grassland merges with oak woodland and chaparral, and these with oak-pine woodland, so you will notice, as you climb steadily higher, that these woodlands gradually mingle with the open pine forests that cover much of the Rincon Mountains above 6,000 feet. PONDEROSA PINE is the "big tree" of the Rincons, usually growing in clear, open stands. Through its high canopy of spreading branches, sunlight mottles the shaded forest floor. Its presence indicates still cooler and wetter conditions than those below. Here you will need blankets at night, though summer days are warm.

Except for grasses such as PINE DROPSEED, SCREWLEAF MUHLY, and MOUNTAIN MUHLY, ground cover is scarce. In tree-glades or on old burns, however, intermediate-type shrubs (such as BUCK-BUSH) and various herbs have established themselves. Some herbs develop into patches of colorful flowers in summer and autumn. Common flowering plants found among the pines are COLOGANIA; PEAVINE, with its large and showy, white, sweetpea-like blossoms throughout the summer; lupines; DOGBANE; and the familiar white WESTERN YARROW. Here, too, may be found GROUNDSEL, ASTER, FLEABANE, and others, often brightened by the presence of butterflies and other insects seeking nectar and pollen. Most of these forest flowers bloom in late summer or autumn, when plants in the desert, far below, are drab and dormant.

Throughout the pine forests, numerous small canyons and rocky outcrops favor the development of thickets of oak and locust, frequently growing together. GAMBEL'S OAK, a leaf-shedding white oak, ranges in size from a small shrub to a handsome tree. It has broad, deeply lobed leaves which provide browse for deer. Its acorns are consumed by deer, rodents, and birds, including wild turkeys. The NEW MEXICAN LOCUST also is browsed by deer. Rarely reaching tree size, this species is an attractive vegetative cover because of its odd-pinnate leaves and large clusters of purplish-pink flowers that appear in May and June. Locusts sprout freely from roots and form expanding thickets which encroach upon oak clumps. They provide a valuable network of soil-holding roots, important in retarding erosion. The best stands occur along the east slopes of the Rincons.

Relatively few in number, compared with the stands of the dominant ponderosa pine, the smaller CHIHUAHUA PINE grows on lower dry slopes and benches. Its needles are shorter than those of the ponderosa pine, and its cones are conspicuously persistent, remaining on the tree for several years. This Mexican species invades the United States in the mountain ranges of southern Arizona and southwestern New Mexico. In the monument it is found mostly in the transition areas between oak-pine woodland and ponderosa pine forest.

The Rincon Mountains are not high enough to provide a fir-forest habitat except in a few favorable locations. On the highest parts of the Rincons, ponderosa pines dominate in the warmer, exposed locations, but whitebarked QUAKING ASPENS grow in pure stands on cooler slopes and with DOUGLAS-FIRS on the north side of Rincon Peak. West of Spud Rock are abundant small groves of MEXICAN WHITE PINE.

A cone-bearing tree growing with Douglas-fir—exclusively on higher northern and northeastern slopes of Mica Mountain—is the WHITE FIR. Flattened, gray-green needles curving upward from the branches, and large, green cones growing upright on limbs near the tops of the trees identify this beautiful evergreen. On open stands, limbs of even the large trees grow from the trunk almost down to the ground. The bark is gray or ash-colored.

BRACKEN forms a green ground cover in heavy stands of pine and fir. This fern grows 3 feet tall over much of the forested Rincon highland. Among the shrubs found on the mountaintop is the SNOWBERRY, whose leaves are browsed by deer and whose berries are eaten by birds and chipmunks.

A spring, a small mountain stream, and a meadow near Manning Camp complete the picture of the higher elevations in the monument. In this bit of meadowland are found BOXELDER, NEW MEXICAN ALDER, CINQUEFOIL, CHOKECHERRY, GOLDENROD, ORANGE SNEEZEWEED, MARIGOLD, WILLOW, and a number of other shrubs, grasses, and herbs characteristic of the high mountain meadows of the Southwest.

Jerusalem cricket (of a different family from the true crickets) has legs adapted for tunneling in sand.

animals
and how they survive

Just as plants depend for their existence on soil, water, and sunlight, so animals, including man, depend on plants. For green plants are the basic food producers in nature, manufacturing carbohydrates, proteins, and other essential compounds from minerals, air, and water, with the help of chlorophyll and the sun's energy. Animals get their food either by eating green plants or by eating animals that have eaten plants. Microscopic decomposers complete this food chain, breaking down dead plants and animals into substances that once again can be used by plants. Since each link in the chain depends on the other links, it's not hard to see that a change in one will cause a change in the others. And because animals depend on plants for cover as well as for food, their fortunes are doubly tied to the welfare of plants.

Animals and plants share some of the same basic problems—particularly, how to stay within tolerable temperature limits, and how to maintain an adequate supply of water. Plants solve these problems mostly by structural adaptations, animals mostly by behavioral. In the desert, for instance, cold-blooded animals such as snakes and lizards (which have no internal control over body temperature) crawl underground or into shade during the midday heat of summer, and come out to hunt food during the cooler hours. Birds and mammals cool themselves through evaporation of water from their bodies. This makes water conservation doubly critical for them; they too handle it by staying in the shade or going underground during hot times. Desert animals get much of their water from the plants and animals they eat, but some species, such as mule deer and Gambel's quail, require large amounts of drinking water as well.

Cold weather poses another problem. Most reptiles and some mammals solve this one by hibernating underground or in rock dens, where temperatures remain moderate throughout the year. Many birds and some mammals migrate to areas where temperatures are warmer and food is more abundant, which may mean going farther south or simply moving down the mountainsides. And insects can survive in a dormant form, as eggs or pupae, though many species remain active during the temperate Sonoran winters.

61

If you want to see animals, then, go where the vegetation is thickest and most varied, and go when temperatures are moderate. During warm seasons in the desert, this means that walking the washes early or late in the day will give you the best chances for seeing wildlife. Coveys of Gambel's quail explode into the air, peccaries snort through the underbrush, butterflies festoon flowering shrubs, and coyotes stealthily hunt.

Invertebrates

Insects are generally not bothered by excessive heat, and many species are active during the hottest hours. This is especially true when the plant blossoming season is at its height. Flowers of the mesquite, paloverde, catclaw, saguaro, and other desert plants are "alive" all through the day, as many species of insects seek nectar and pollen or prey on other insects attracted to the blossoms. Insects are fed upon by various species of birds; flycatchers flock to parts of the desert where nectar-yielding flowers are numerous. Because of the absence of extreme cold, the desert climate enables insects to be active throughout much of the year and to support a considerable bird population.

Insects play a far more important role in the plant and animal life of the desert than is usually realized. Many desert flowers must be insect pollinated to produce viable seeds. Birds of many kinds depend upon insects for food, and even the seed-eating birds, during the nesting season, rely upon insects to provide the enormous quantities of food and moisture required by their fast-growing nestlings. Many other desert creatures, including certain

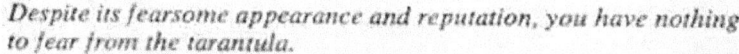

Despite its fearsome appearance and reputation, you have nothing to fear from the tarantula.

62

snakes and lizards and some spiders, depend upon insects for food. The body juices of the insects provide the all-important moisture—which these creatures can get from no other source. Bats, too, are insect eaters, spending the hours of darkness in seemingly aimless and erratic flight while foraging for moths and other night-flying insects that visit the light-colored blossoms of night-blooming plants.

Some species of insects may become so numerous that they threaten the very life of the plants on which they live. Pine bark beetles annually damage or kill numbers of pinyons and ponderosas in the Rincon Mountains, but have been kept sufficiently under control by their natural enemies so that their ravages have not reached epidemic proportions.

Among the common spectacular insects is the TARANTULA HAWK, a large blue-black, red-winged wasp that preys on large spiders. Temporarily paralyzing the spider with its sting, the wasp lays a single egg on its victim, thereby assuring an abundance of living food for its young. The PRAYING MANTIS is another large insect, usually green and inconspicuous among the foliage of desert plants, which it frequents in search of small insects. Ants of many species are active almost everywhere in the desert, harvesting seeds of various plants. Some species construct mazes of underground nest tunnels and deposit the excavated materials on the surface, forming conical, sometimes, craterlike, anthills.

Along with the insects, other arthropods (jointed-leg creatures with exoskeletons) find their home in the desert. The arachnids (eight-legged arthropods) include spiders and scorpions. Of the

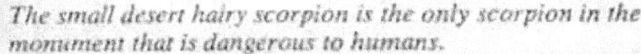

The small desert hairy scorpion is the only scorpion in the monument that is dangerous to humans.

The Colorado River toad is exceeded in size among U.S. toads only by the giant toad of south Texas (and Mexico).

former, the NORTH AMERICAN TARANTULAS are famous for their large size and formidable appearance, which have given them the wholly undeserved reputation of being dangerous to humans. The really dangerous creatures are the SCORPIONS, whose long, flexible tails bear a poisonous stinger at the tip. Several species are found in the monument; but only the small, straw-colored scorpion has venom known to have been fatal to humans. The other scorpions found in the area can inflict painful stings, but with only localized and rarely serious effects.

Amphibians

As might be expected, amphibians are scarce in the monument because of lack of permanent water. The few springs and seeps, however, furnish excellent breeding places for several species of amphibians. Best known among these are the RED-SPOTTED TOAD, LEOPARD FROG, and CANYON TREEFROG, the latter common near Manning Camp. A spectacular desert amphibian and the second largest toad in the United States, is the huge COLORADO RIVER TOAD, sometimes found near residences in the evening when outdoor lights attract swarms of insects.

Reptiles

Except for small lizards, reptiles are not much in evidence in the monument. Nevertheless, they are present and are important in

*The short-horned lizard, more cold-tolerant than the desert horned
lizard, ranges from the foothills into the mountain forests of Saguaro.*

the various plant-and-animal communities in which they live.
Almost all lizards are insectivorous, and along with birds and other
insect-eaters help to keep the number of insects within bounds.
A notable exception is the GILA (HEE-lah) MONSTER. (See appendix for scientific names of reptiles) largest of the lizards found
in the United States. (It is one of the world's two poisonous lizards,
and the only one found in this country.) The gila monster is
especially fond of bird eggs, and also eats nestlings and small
rodents, obtaining necessary moisture from their body juices.
These food habits are quite similar to those of the several species
of snakes found in the monument, the majority of which are per-
fectly harmless to humans.

Just as the lizards help to control the insect population, the
snakes are important in preventing the buildup of large numbers
of rodents that would result in widespread damage to vegetation.
Visitors to the monument rarely have the opportunity to observe
snakes, since they are in hiberation during winter and remain in
the shade or in underground burrows during the hot part of each
summer day. Perhaps those most frequently seen are the GOPHER
SNAKE and the COACHWHIP. Many desert snakes hunt only at
night; others that are normally active during days of moderate
temperatures become night hunters during hot weather. Although
they are not abundant, there are several kinds of rattlesnakes in
the monument, the commonest desert species being the WESTERN

DIAMONDBACK and MOHAVE RATTLESNAKES. Except for the small, very rare, and secretive ARIZONA CORAL SNAKE, rattlesnakes are the only poisonous snakes in the monument. Snakes, like other living things in the national monument, are protected by law.

Don't be surprised while following a desert footpath to come upon a plodding tortoise. This bona fide desert dweller, the DESERT TORTOISE, is a vegetarian, feeding on cactus, grass, and other low-growing plants.

Birds

Because of its great variety of habitats, ranging from luxurious desert vegetation to deep mountain forests, Saguaro National Monument provides food and shelter for many species of birds. Some of these reside all year within a single zone, while others move upward in summer, returning to the desert when the mountaintops are covered with snow. Other species pass through the monument in spring and autumn in their annual migrations between Latin America and summer nesting grounds in northern United States or Canada. The following species are those you are most likely to see.

A common year-round desert resident is the CURVE-BILLED THRASHER, noticeable, noisy, and about the size of a robin. These

The Mojave rattlesnake prefers desert, grassland, and open brushland to densely vegetated areas.

energetic inhabitants of the cactus forests nest in mesquite clumps and cholla thickets. Their diet—they feed on insects and succulent fruits—makes them comparatively independent of water. The thrashers are delightful songsters. The CACTUS WREN, looking somewhat like a small thrasher, is even noisier. It protects its nest by building in a cactus. This wren lives largely on a diet of insects, but about 17 percent of its food is seeds and fruits. One of the most attractive of the ground birds is the GAMBEL'S QUAIL. Many coveys are found throughout the desert within close range of water. In winter, Gambel's quail feed mainly on seeds, berries, and plant shoots; in summer they augment this diet with ants, grasshoppers, and other insects. The ROADRUNNER, ungainly clown of the desert, is frequently seen by visitors as it scuttles through the underbrush along the margins of sandy washes. Not particular as to food, it is sometimes seen with the tail of a lizard protruding from its beak, and it is known to eat insects and spiders, snakes, young rodents, small birds, seeds, and fruits. Quite capable of flight, the roadrunner prefers to trust to its legs and the security of thickets, but will take to its wings if pursued in the open.

Two members of the woodpecker family closely associated with the saguaro cactus are the GILA WOODPECKER and GILDED FLICKER. Called carpenter birds because they drill nest holes or pockets in

You may mistake the curve-billed thrasher's call, "whit-wheet," for a human whistle of attention.

The gilded flicker drills its nest hole in the saguaro.

the saguaro stems, both species are of particular interest because of their limited range and specialized habitat. Two other desert birds, sufficiently similar to be confused, are the red, black-faced CARDINAL and the gray-and-red PYRRHULOXIA, both of which have crests. Look for these species in the shrubs along washes. Several kinds of doves are found in the desert, including the common MOURNING DOVE and the WHITE-WINGED DOVE. Mourning doves are all-year residents, while the large white-wings drift in from Mexico in May, remaining long enough to raise families and join other animals in harvesting fruits of the saguaro.

Seen and heard in the desert all year, the canary-voiced HOUSE FINCH raises its family among cholla and mesquite thickets. The tinkling song of the ROCK WREN is a familiar sound in the desert in winter. These gray ground dwellers go farther north or to higher elevations to nest.

The PHAINOPEPLA is one of the most noticeable of the desert birds because of its silky crest, glossy black plumage, and habit of perching on the topmost branch of a tree while indulging in flutelike song. A permanent resident of the monument, subsisting on mistletoe berries and other vegetable matter in winter, it has a diet of insects, principally ants, during the rest of the year. Flycatchers are especially abundant and conspicuous during spring

68

The white-winged dove's interest in the saguaro is in the nectar and fruit.

and early summer when the blossoms of trees, shrubs, and the larger cactuses attract swarms of insects. Among these birds are SAY'S PHOEBE and ASH-THROATED FLYCATCHER. The LESSER NIGHTHAWK lives on a diet of insects, which it catches while on the wing. It is especially noticeable from May to September as it skims the tops of the tallest saguaros in the dusk of evening. The lesser nighthawk also ranges up to the oak woodlands.

Predators are an integral part of the bird population, one of the smallest and most active being the LOGGERHEAD SHRIKE. This black-and-gray bird gorges itself on beetles and grasshoppers when insects are abundant, turning to lizards, rodents, and small birds at other times. It has the unusual habit of impaling its prey on thorns for future use. The RED-TAILED HAWK is the commonest of the large soaring hawks, which live mainly on rodents and reptiles. It builds its large stick nest in the forks of saguaro arms. Like the shrike and the SPARROW HAWK, the red-tailed hawk is found in grasslands, chaparral, and woodlands as well as in the desert. Because of their nocturnal habits, owls are not often seen by visitors, but they are abundant in the monument. In addition to the GREAT HORNED OWL, which like the red-tailed hawk feeds principally on rodents and builds cumbersome nests in saguaro branches, the SCREECH OWL and the tiny ELF OWL are numerous

69

in the cactus community. Screech and elf owls make use of abandoned woodpecker holes in saguaros, not so much for nesting as for dark and comfortable hiding places during daylight hours; they emerge after sunset to hunt insects and small rodents. Best known of the carrion eaters, the TURKEY VULTURE is rarely seen on the ground, but is a common sight, singly or in groups, circling high in the sky.

The oak-pine-juniper woodland has its set of birds too. One of the noisiest, most quarrelsome, and most mischievous is the MEXICAN JAY, a permanent resident. In summer, it shares this habitat with the night-flying POOR-WILL, which closely resembles the nighthawk but lacks the white wing patches. Shy, secretive, and protectively colored, this bird is rarely seen, but its plaintive call is a familiar twilight sound at the middle elevations of the mountains. Here, too, is found the strikingly patterned HARLE-QUIN QUAIL, which waits until you are almost upon it before flushing. The RUFOUS-SIDED TOWHEE prefers brushy slopes and canyons, where it trills its monotonous song from the branch of

The red-tailed hawk builds its nest in the fork of a saguaro and by day ranges over the entire monument in search of prey.

a skunkbush or scratches noisily and industriously among the fallen leaves beneath an oak. And anywhere from the oak-pine woodland to the top of the Rincons, you are likely to startle the large BAND-TAILED PIGEON from its perch.

The pine and fir forests of the higher Rincons are rich in bird life in summer, but only a few species remain the year around. Among the latter is the STELLER'S JAY, which includes insects in its summer diet but feeds mainly on acorns and other vegetable material in winter. The MOUNTAIN CHICKADEE is busy among the branches of the large pines and firs even when snow lies deep on the ground. PIGMY NUTHATCHES—tiny, constantly twittering insect-eaters—also remain all year.

In summer, when insects are plentiful, the mountain forests are alive with birds. The MEXICAN JUNCO, which winters at lower elevations, hides its nest in grass clumps among the firs and aspens. Hummingbirds of several species pause on vibrating wings to suck nectar from blossoms in the meadow at Manning Camp. Brightly

The great horned owl, a nocturnal counterpart of the red-tailed hawk, nests in the saguaro and feeds primarily on rodents (left). Steller's jay lives year-round in the pine and fir forests of the higher Rincons.

colored summer visitors such as the HEPATIC TANAGER and the PAINTED REDSTART search for insects among the pine boughs or flash in the sunlight as they flutter across open glades in the forest. The mountaintops, too, have their predators; the COOPER'S HAWK, which nests in wooded canyons, is large enough to lift a pigeon or rabbit, but generally preys on rodents and smaller birds. The largest bird known to inhabit the Rincon Mountains is the TURKEY, which nests and raises its young among the firs and aspens. It descends in winter to the oak-pine woodlands, where it feeds on pinyon nuts, acorns, and grass seeds.

Mammals

The most frequently seen mammals in the monument are rodents and members of the hare clan. Among the latter group are two rabbits. The DESERT COTTONTAIL is common in the lower levels of the desert, and the EASTERN COTTONTAIL inhabits the mountains to 8,600 feet. Adaptable to a wide range of environmental conditions, these animals augment the water they obtain from springs with moisture derived from sap. There are two species of JACKRABBITS (actually hares) in the monument. They remain at the lower levels, where they are a common sight amid the cactuses. Unlike rabbits, which are born naked, blind, and helpless, hares are born with fur, open eyes, and the ability to move about.

Among the monument rodents the largest are the PORCUPINES.

The Yuma (or gray-tailed) antelope ground squirrel, a chipmunk-like rodent of the desert and grassland (left). In the monument, the cliff chipmunk prefers the oak-pine woodland and the coniferous forest.

Though rarely seen, they leave characteristic scars on pinyons and ponderosa pines, recording their feeding habits at higher elevations. They are also active in the chaparral belt, and an individual occasionally wanders down into the desert where it eats mesquite beans and samples the bark of ocotillos and other shrubs and trees.

Several species of ground squirrels are abundant. At the lower levels, especially among creosotebushes, the ROUNDTAIL GROUND SQUIRREL finds suitable living conditions, while the YUMA ANTELOPE GROUND SQUIRREL ranges from the cactus forests into grassland. At this elevation and upward through the pinyons and junipers, the ROCK SQUIRREL makes its burrows in rocky ledges and brushy canyons. CLIFF CHIPMUNKS enliven the oak-pine woodland and higher forests with their quick movements and cheerful chatter.

KANGAROO RATS honeycomb the soil from the cactus forests up to the oak-pine belt. Remaining underground during the day, they are frequently seen at night. These animals do not require free water; they obtain adequate moisture from a chemical process within the body during the digestion of food, which is mostly dry seeds. WOODRATS, famous in song and story as pack rats or trade rats because of their habit of carrying away objects of human use and leaving something else in their place, are found throughout the monument at all elevations. Look for their stick nests among the pads of prickly pear, a favorite food.

Other rodents common in the monument include the CACTUS MOUSE, GRASSHOPPER MOUSE, DEER MOUSE, and VALLEY POCKET GOPHER.

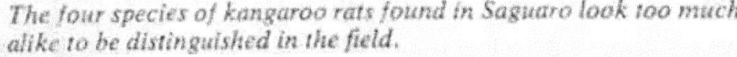

The four species of kangaroo rats found in Saguaro look too much alike to be distinguished in the field.

Mule deer.

Among the larger mammals in the monument are two species of deer. The MULE DEER subsists in winter on cactus fruits, ephemerals, and other desert vegetation. In summer they find abundant browse in the higher oak woodland. The forested areas along the crest of the Rincons support a population of the smaller WHITE-TAIL DEER. These graceful animals browse on aspen, buckbrush, and other shrubs and small trees. They are particularly fond of acorns. When snow flies some descend to the protective cover of the oak-pine woodlands and chaparral.

PECCARIES, characteristic of southwestern deserts, usually travel in herds of from three to as many as 50 animals. They wander through the groves of mesquite along desert washes, and root among beds of pricklypear. Pricklypear pads are their chief food; they are said to feed upon cactus fruits in summer and autumn. In addition to the moisture obtained from succulent stems and fruits, peccaries require considerable water, hence they frequent springs and seeps. Small bands of these animals occasionally visit the waterhole near the visitor center, where they are excitedly watched by visitors fortunate enough to be in the lobby at the time.

Saguaro National Monument also has a large number of predatory animals. Many of these, popularly believed to be exclusively meat-eaters, actually also eat much vegetable matter. The predators—an exciting part of the monument's fauna—play an important role in preventing overpopulation of the prolific rodents.

Chief among predators is the COYOTE, which ranges throughout the monument. In winter, coyotes are found principally below

74

Collared peccary, or javelina.

6,000 feet, where hunting is easier and where rodents remain longer out of hibernation. They are also known to roam the forested heights of the Rincons and Tanque Verde Ridge. Studies of coyotes in the monument made by biologist Lowell Sumner in January 1951 showed that their winter diet consisted of about 78 percent fruits and seeds, 11 percent small rodents, 7 percent deer, 4 percent birds, reptiles, insects, and carrion, and 1 percent debris. The coyote is one of the principal wild creatures associated with the history of the West, and its continuing presence in the monument brings a thrill of pleasure to visitors who hear the voices of the pack during the calm of evening, or catch a glimpse of one of these gray canines trotting through open stands of saguaros. BOBCATS, ranging over much the same territory as coyotes, subsist principally on rodents, birds, and insects. Because of their secretive habits, they are seldom seen.

The GRAY FOX is another fairly common inhabitant of the cactus desert and higher brushlands, and is also known to roam the forested uplands. It is usually seen at night. Rodents are its principal source of food, but it also preys on birds and reptiles and eats much vegetable matter. The smaller and rather rare KIT FOX, restricted mostly to the desert, is also a night hunter. Kangaroo rats are a favorite item in the kit fox diet, but these small predators also eat grasshoppers and other insects.

Skunks, members of the weasel family, are relatively common in the monument, and several species inhabit areas where water is available. They are usually active at night. The HOGNOSE SKUNK is a desert dweller recognizable by its solid-white back. It roots for insect larvae and eats cactus fruits, bird eggs, and nestlings. The STRIPED SKUNK ranges throughout the monument; the SPOTTED SKUNK is found at all elevations, usually among rocks. Insects, rodents, and fruits are its main food.

Another member of the weasel family, the BADGER, is occasionally seen in the desert although it is by no means limited to that habitat. The badger feeds mainly on rodents, which it digs from burrows with its strong forelegs and heavy claws.

The RACCOON, longer-legged than the badger, is readily recognized by its gray fur, black mask, and ambling gait. It inhabits brushy canyons having permanent water, and sometimes wanders up into the pine belt in summer. Amphibians, scarce as they are, and other water creatures are among the preferred foods of raccoons; but they also eat small rodents and plant food, including berries, acorns, and other fruits.

The RINGTAIL, a smaller relative of the raccoon, is somewhat similar in habitat preference and nocturnal habits. Its flattened, bushy tail, acting as a balancer, helps this short-legged, agile animal in leaping from point to point on the steep rocky surfaces it seems to favor. It sometimes takes up residence in little-used or abandoned buildings, where small rodents, its principal

Badger.

source of food, are usually abundant. About the size of a house cat, it has large eyes and ears and alternating dark and light bands on its tail.

A tropical animal that seems to be extending its range northward, the COATI (or coatimundi) is often seen in the forests of the Rincons. With long snout and long, banded tail, it looks something like an elongated raccoon, to which it too is related These omnivorous animals travel in bands, rooting among leaves for insects and whatever else they can find.

Coati, or coatimundi.

the rhythms of nature

Natural landscapes may appear unchanging, but this is illusion. Within the apparent constancy, daily and seasonal cycles, fluctuations in numbers, and long-term change are the rule.

Daily cycles are obvious to those who are about at the edges of the day. Take 24 summer hours in the cactus forest of Saguaro National Monument. When the first light comes over the mountains, curve-billed thrashers and cactus wrens sing noisily among the chollas. Other birds soon join in. The early morning walker is likely to hear peccaries grunting in the mesquites along a wash, or see mule deer staring at him, frozen like statues before sudden flight to a sheltering thicket.

At midday, the scene is quiet. Nothing stirs under the baking sun except perhaps a vulture, soaring on the hot air currents. The desert creatures have not gone—they are in the shade of bushes or underground. Even some of the plants are "taking a siesta," having folded their leaves or closed their leaf pores.

Soon after sundown the desert comes to life again. The birds give a subdued version of their morning's vocal performance. Tarantulas begin their slow, stately walk over the ground searching for prey or mates. Coyotes stretch and howl—a prelude to the evening's hunt. As night falls, rattlesnakes emerge from their cool retreats to search out kangaroo rats, which in great numbers are scrutinizing the sand for seeds. And through the night, creatures of many other kinds hunt food to last them through another broiling day.

The rhythmic patterns of the daily cycle are paralleled on a larger scale through the year—seasons of activity follow seasons of quiet. In the desert, rain or lack of rain marks the changes, though gradually rising or falling temperature adds its impact.

The gentle rains of winter prepare the way for the year's greatest burst of activity. By March, spring flowers are blooming and birds are starting to nest. Snakes begin to come out of hibernation. April and May see the apex of spring activity, as insects swarm around the flowering plants, and birds take advantage of this proliferation of food to raise their young. The desert now is

Desert vegetation near Red Hill, Tucson Mountain Section.

yellow with the blossoms of paloverde, mesquite, acacia, and brittlebush.

But April also marks the beginning of a drought that intensifies through May and June, making these last 2 months the year's parching crucible in which reproductive ability is tested. If winter rains have been meager, the heat and drought of May and June can kill all the young of many birds. Some birds, such as Gambel's quail, may not even attempt to nest in a dry year. Conditions may be so harsh at this season that some mammals, such as the pocket mouse, close up shop completely, sleeping the days away underground.

Relief comes with the rains of July and August. Now the summer annuals spring magically from the ground, perennials put forth new leaves, and saguaros do all their growing for the year. This summer burst of plant growth is accompanied by a new hatching of insects, which allows a few more birds to nest, and along with the new vegetation supports a larger pyramid of animal life generally. Among the new animals that reappear are toads, which now emerge from their long sleep in the soil to mate and lay eggs in the pools formed by summer rains.

When the last torrential rain of August or September falls, a new dip in the yearly cycle of activity begins. This one is not so deep, not so trying, as the drought of early summer, but it too is a time of relative quiet. Roundtail squirrels go underground to sleep until cooler weather comes. Now the migrating birds slip through, hardly noticed among the mesquites and paloverdes. Butterflies lay their eggs, in preparation for a new generation beyond the winter. Signaling the last phase in the yearly cycle, wet canyons turn yellow and brown as cottonwoods, willows, and sycamores present a pale version of the spectacular foliage displays seen in the East.

While these daily and seasonal cycles are following their well-known courses, each species of plant and animal is undergoing its own fluctuations, in a constant struggle that generally goes unnoticed. For the balance of nature is not a static one, but more like the rocking of a seesaw on its fulcrum. The population of a species goes up one year, down another—depending on the weather, the food supply, predators, competitors, and a thousand interactions that reverberate through the community in which it lives. The numbers of some species, like the Gambel's quail, fluctuate wildly from year to year, while those of others, such as the harvester ant, remain quite stable. But the oscillations of the seesaw, big and little, average out from year to year so that the species maintains itself in the community. The other members are going through the same thing, in a system of checks and balances that over the short run keeps the whole community nearly constant.

But over decades, centuries, or longer, the fulcrum of the seesaw moves: the larger environment changes, and the community and its constituent plants and animals must change with it or perish. Such changes may be climatic, as we saw with the formation of the Southwestern deserts; or it may be geologic, as with the rising and partial disintegration of the Rincon Mountains. The efforts of plant and animal species to meet such changes constitute in large part the story of evolution, for new environments spawn newly evolved forms of life. Evidence of such evolution we have seen in the plants of the Sonoran Desert, notably the giant saguaro cactus, whose prickly ancestor lived in the West Indies only 20,000 years ago.

Thus nature is ever-changing; and the inexorable rule for all living things is, "adapt or perish." Before technological man enters the scene, the slow evolutionary process can keep pace with the changing environments, though here and there a species is dropped by the wayside. Generally, communities of living things reach new equilibria without serious disruption.

But what happens when man, with his machines and his passion for progress, institutes changes of a speed and kind and on a scale drastically different from those brought about by earthquakes, storms, shifting climates, and other natural phenomena? What happens to the living things that have adapted to the harsh desert environment when that environment is drastically altered?

the impact of man

Man has been a part of the scene in this region for several thousand years, but until recent times his influence on it was minimal. Only with the rapid technological development of the last century has man been able to make a major impact on this landscape. Thus the story of man, here as elsewhere, is a story of gradually accelerating power to change environments, a power that now threatens to destroy environments, and with them, man himself.

From carbon-14 dating in Ventana Cave, we know that man was here at least 12,500 years ago, in the Pleistocene age, a time that was cool and moist compared to the present. Living by hunting, he followed mammoths and other large mammals. As the climate warmed during succeeding millenniums, and these mammals became extinct, he came to rely more on plant foods. These hunters and gatherers necessarily had to live in small bands scattered over the land, since the plants and animals on which they depended were widely dispersed. By 300 B.C., they had learned from people to the south how to cultivate food plants, and had developed a sedentary way of life. About 2,300 years ago a group we call the Hohokam settled in the Salt and Gila River basins (including the Santa Cruz Valley). By A.D. 700 they had a well developed agricultural economy including extensive irrigation systems. Pottery fragments, projectile points, petroglyphs (rock carvings), and other evidence show that Hohokam villages existed for about 600 years in the eastern section of the monument along Rincon Creek and its tributary washes. Archeological work in the Tucson Mountain Section has indicated that this area was visited only temporarily by the Hohokam, for hunting, food gathering, and perhaps ceremonial purposes.

During the 15th century the Hohokam high culture vanished. Soils made salty from irrigation water and internecine warfare are suggested explanations.

When the Spanish explorer Coronado passed to the east of the Rincons in 1540, he found the Sobaipuri living there. The Pimas, descendants of the Hohokam, occupied the same basins the Hohokam had. To the west, in drier country, lived the Papago. These tribes, thought to be descendants of the Hohokam, lived much the same sort of life, practicing irrigation where surface water was available, hunting and gathering where it was not.

83

The period of Spanish rule, implemented by a series of missions, began in the Santa Cruz Valley about 1692, when the energetic Father Kino began his work among the Pima and Papago. The mission system concentrated the Indians in fewer places, brought Spanish and, later, Mexican settlers into southern Arizona, and introduced sheep, cattle, and goats. Although the new culture must have had some environmental effects, there is no evidence of drastic change. Grass was plentiful, and streams, including the Santa Cruz, remained marshy and unchanneled.

After the Gadsden Purchase of 1853-54, however, when the present boundary with Mexico was established and this area came into United States ownership, man's impact on the land increased. Apache raiding had been a deterrent to settlement during the 18th and 19th centuries, but, after the Civil War, American soldiers got the upper hand and settlement increased. Following completion of the Southern Pacific Railroad to Tucson in 1879, a cattle boom began. The disastrous results of the livestock explosion of the eighties—overgrazing, soil erosion, and starvation of cattle— we have already seen in the story of the saguaro cactus. In 1890, a flood cut a deep channel in the Santa Cruz River, transforming it from a meandering, marshy stream to the usually dry incision one sees today. The arroyo cutting of this and many other rivers throughout the Southwest was undoubtedly due partly to increasing aridity, which reduced the plant cover and its water-holding capacity. But the erosion was probably triggered by overgrazing.

In the monument, we have already seen how grazing pressure, hunting, and predator control reduced ground cover and led to an upsurge of certain rodents and a decline in large mammals. But there have been other man-induced changes. For as long as there has been forest on top of the Rincons, there has been fire. Lightning-caused fire is a natural part of ponderosa-pine forest, every few years burning the litter and small trees and shrubs from the forest floor, and thus maintaining open stands of tall trees. But since 1908, when the Rincon Mountains came under protection of the Forest Service, U.S. Department of Agriculture (to be followed in 1933 by National Park Service, U.S. Department of the Interior protection), fires have been put out as fast as possible. This policy has resulted in a paradox. On the one hand, thickets of scrawny young pines and shrubs such as buckbrush have developed in many places under the tall pines. On the other hand, the accumulation of litter and low-level vegetation has provided fuel over the years for occasional very hot crown fires, which have been hard to control and which have burned large acreages. On top of the Rincons you can see several meadows that resulted from these fires. Only a few scattered trees and stumps remain in them to suggest the forest that once was there.

Ideally, national parks and monuments should be "vignettes of primitive America"—naturally evolved landscapes in much the same condition as when first seen by Europeans. In reality they are compromises—beautiful, wild, but still bearing the marks of human occupation. In Saguaro, as we have seen, fire control has produced a forest different from that known to the Indians who once lived here; grazing has depleted the ground cover; and hunting has removed the desert bighorn from its rocky haunts. In these days of burgeoning population, when human influence is affecting every natural landscape, environmental management becomes necessary to approach the ideal of naturalness. This may mean "prescribed burns" to return forests to their earlier state; elimination of grazing; or reintroduction of animals once native to a park. In the summer of 1971, after 2 inches of rainfall, natural burns (caused by lightning strikes) were allowed to run their courses.

Some or all of these measures may be taken in Saguaro, in order that future generations will know a piece of the Sonoran Desert as it was in Coronado's time.

The realization of this goal, however diligently we work toward it, seems almost each day to become more difficult of attainment. These desert and mountain environments—which once seemed secure, needing only the continued protection afforded by their status as a national monument—are increasingly imperiled by the works of man. As the city of Tucson sprawls in all directions, the monument's two divisions, islands in an encroaching sea of civilization, must withstand ever-accelerating hazards. Vandalism takes an increasing toll of the saguaros; housing developments creep toward the monument borders. Smog drifts over the fragile plant communities, threatening to choke them—as the polluted air from Los Angeles is already strangling forests in the distant San Bernardino Mountains.

A new awareness that the best-managed preserve cannot thrive independently of what is happening in the surrounding region only emphasizes the difficulty of the task. Saving the saguaros is inevitably tied to the problem of enhancing the quality of life and reversing the degradation of the environment—not only in Tucson but throughout the Southwest.

There is no time to waste. Only concerted effort by scientists, resource managers, and the community can assure that our grandchildren will be able to visit a Saguaro National Monument where coyotes howl under the moon, peccaries snort through the washes, and giant cactuses lift bristly green arms into a blue sky.

appendix

Suggested Reading

Arnberger, Leslie P. *Flowers of the Southwest Mountains*. Southwestern Monuments Association, Popular Series No. 7. Globe, Ariz. 1962.

Benson, Lyman D. *The Cacti of Arizona*. University of Arizona Press, Tucson. 1969.

Burns, William A. (ed.). *The Natural History of the Southwest*. Franklin Watts, Inc., New York. 1960.

Dodge, Natt N. *Flowers of the Southwest Deserts*. Southwestern Monuments Association. Popular Series No. 4. Globe, Ariz. 1961.

————. *Poisonous Dwellers of the Desert*. Southwestern Monuments Association. Popular Series No. 3, Globe, Ariz. 1964.

Dodge, Natt N., and Herbert S. Zim. *The Southwest*. Golden Press, New York. 1962.

Earle, W. Hubert. *Cacti of the Southwest*. Desert Botanical Garden. Phoenix, Ariz. 1963.

Jaeger, Edmund C. *The North American Deserts*. Stanford University Press, Stanford, Calif. 1957.

Kearney, Thomas H., Robert H. Peebles, and collaborators. *Arizona Flora*. University of California Press, Berkeley. 1960.

Krutch, Joseph W. *The Voice of the Desert*. William Sloane Associates, Inc., New York. 1955.

Milne, Lorus, and Margery Milne. *The Balance of Nature*. Alfred A. Knopf, Inc., New York. 1960.

Olin, George. *Mammals of the Southwest Deserts*. Southwestern Monuments Association, Popular Series No. 8. Globe, Ariz. 1965.

————. *Mammals of the Southwest Mountains and Mesas*. Southwestern Monuments Association, Popular Series No. 9. Globe, Ariz. 1961.

Patraw, Pauline M. *Flowers of the Southwest Mesas.* Southwestern Monuments Association. Popular Series No. 5. Globe, Ariz. 1959.

Phillips, Allen, Joe Marshall, and Gale Monson. *The Birds of Arizona.* University of Arizona Press, Tucson. 1964.

Schmidt-Nielsen, Knut. *Desert Animals: Physiological Problems of Heat and Water.* Oxford University Press, Inc., New York. 1964

Sears, Paul B. *Deserts on the March.* University of Oklahoma Press, Norman. 1959.

Stebbins, Robert C. *Amphibians and Reptiles of Western North America.* McGraw-Hill Book Co., New York. 1954.

Storer, John H. *The Web of Life.* The Devin-Adair Co., Old Greenwich, Conn. 1960.

Sutton, Ann, and Myron Sutton. *The Life of the Desert.* McGraw-Hill Book Co., New York. 1966.

Underhill, Ruth. *People of the Crimson Evening.* Publications Service, Haskell Institute, Lawrence, Kans. 1951.

Common and Scientific Names of Plants

Agave—*See* Amole
Alligator juniper—*Juniperus deppeana*
Amole—*Agave schottii*
Arizona cypress—*Cupressus arizonica*
Arizona rosewood—*Vauquelinia californica*
Arizona sycamore—*Platanus wrightii*
Arizona white oak—*Quercus arizonica*
Aster—*Aster commutatus*
Barrel cactus—*Ferocactus wislizenii, F. lecontii*
Beargrass—*Nolina microcarpa*
Bladder-pod—*Lesquerella gordoni*
Bluedicks—*Dichelostemma pulchellum*
Blue paloverde—*Cercidium floridium*
Bluegrass—*Poa fendleriana*
Boxelder—*Acer negundo*
Bracken—*Pteridium aquilinum*
Brittlebush—*Encelia farinosa*
Buckbrush—*Ceanothus fendleri*
Bullgrass—*Muhlenbergia emersleyi*
Catclaw—*Acacia greggii*
Chain fruit cholla—*Opuntia fulgida*
Chihuahua pine—*Pinus chihuahuana*
Chokecherry—*Prunus virginiana*
Cinquefoil—*Potentilla subviscosa*
Cologania—*Cologania lemmoni*
Creosotebush—*Larrea tridentata*
Curly mesquitegrass—*Hilaria belangeri*
Desert chicory—*Rafinesquia neomexicana*
Desert Christmas cactus—*Opuntia leptocaulis*
Desert-marigold—*Baileya multiradiata*
Dogbane—*Apocynum androsaemifolium*
Douglas-fir—*Pseudotsuga menziesii*
Emory oak—*Quercus emoryi*
Fairy-duster—*Calliandra eriophylla*
Fiddleneck—*Amsinckia intermedia*
Filaree—*Erodium cicutarium*
Fishhook cactus—*Mammillaria sp.*

Fleabane—*Erigeron arizonicus*

Gambel's oak—*Quercus gambelii*

Goldenrod—*Solidago sparsiflora*

Goldfern—*Pityrogramma triangularis*

Gourd—*Cucurbita digitata*

Grama—*Bouteloua sp.*

Groundsel—*Senecio neomexicanus*

Hairy grama—*Bouteloua hirsuta*

Hedgehog cactus—*Echinocereus sp.*

Houstonia—*Houstonia wrightii*

Indian wheat—*Plantago purschii*

Indigobush—*Dalea sp.*

Ironwood—*Olneya tesota*

Jointfir—*Ephedra sp.*

Jojoba—*Simmondsia chinensis*

Lupine—*Lupinus sp.*

Marigold—*Tagetes lemmoni*

Mesquite—*Prosopis juliflora*

Mexican blue oak—*Quercus oblongifolia*

Mexican pinyon pine—*Pinus cembroides*

Mexican white pine—*Pinus strobiforms*

Mock-pennyroyal—*Hedeoma hyssopifolium*

Mountain-mahogany—*Cercocarpus breviflorus*

Mountain muhly—*Muhlenbergia montana*

Netleaf hackberry—*Celtis reticulata*

New Mexican alder—*Alnus oblongifolia*

New Mexican locust—*Robinia neomexicana*

Nightblooming cereus—*Peniocereus greggii*

Ocotillo—*Fouquieria splendens*

Orange sneezeweed—*Helenium hoopesii*

Palmer oak—*Quercus palmeri*

Paper flower—*Psilostrophe cooperi*

Parry's penstemon—*Penstemon parryi*

Peavine—*Lathyrus graminifolius*

Pencil cholla—*Opuntia arbuscula*

Phacelia—*Phacelia crenulata*

Pincushion cactus—*Mammillaria sp.*

Pine dropseed—*Blepharoneuron tricholepis*

Pointleaf manzanita—*Arctostaphylos pungens*

Ponderosa pine—*Pinus ponderosa*

Pricklypear—*Opuntina engelmannii* and others

Puccoon—*Lithospermum multiflorum*

Quaking aspen—*Populus tremuloides*

Sacahuista—*See* Beargrass

Saguaro—*Carnegiea gigantea*

Scorpionweed—*Phacelia crenulata*

Screwleaf muhly—*Muhlenbergia virescens*

Shindagger—*See* Amole

Shrub live oak—*Quercus turbinella*

Sideoats grama—*Bouteloua curtipendula*

Silktassel—*Garrya wrightii*

Silverleaf oak—*Quercus hypoleucoides*

Skunkbush—*Rhus trilobata*

Snakeweed—*Futierrezia sp.*

Snowberry—*Symphoricarpos oreophilus*

Sotol—*Dasylirion wheeleri*

Spanish dagger—*Yucca schottii*

Sprucetop grama—*Bouteloua chondrosioides*

Staghorn cholla—*Opuntia versicolor*

Tanglehead—*Heteropogon contortus*

Teddy bear cholla—*Opuntia bigelovii*

Texas bluestem—*Andropogon cirratus*

Triangle bursage—*Franseria deltoidea*

Turpentine-bush—*Haplopappus laricifolius*

Vine mesquite grass—*Panicum bulbosum*

Western yarrow—*Achillea lanulosa*

White fir—*Abies concolor*

White tackstem—*Calycoseris wrightii*

Wild carrot—*Daucus pusillus*

Wild-cucumber—*Marah gilensis*

Wild-heliotrope—*Heliotropium curassavicum*

Willow—*Salix* sp.

Wolfberry—*Lycium* sp.

Wolftail—*Lycurus phleoides*

Yellow paloverde—*Cercidium microphyllum*

Reptiles and Amphibians of the monument.

This checklist names reptiles and amphibians that have been seen, or that should occur, according to range maps and distribution records in important reference works, in the monument. Four of these species are found only in the Tucson Mountain Section; they are: desert iguana, desert horned lizard, western shovel-nosed snake, and sidewinder. An asterisk marks those species most commonly seen in the daytime.

The *desert* (D) habitat is the rather flat or gently rolling terrain below 3,200 feet in elevation, as seen in the vicinity of the visitor center and the Cactus Forest Drive, and in the north and west portions of the Tucson Mountain Section.

The *foothills* (F) habitat includes the area above 3,200 feet where the land becomes quite rocky and begins to ascend rather sharply on the Cactus Forest Drive and along Hohokam Road in the Tucson Mountain Section.

The *mountain* (M) habitat is restricted to the Rincon Mountains above an elevation of 6,500 feet where one finds tall trees.

	common name	scientific name	habitat
AMPHIBIANS	*toads and frogs*		
	Colorado River Toad	*Bufo alvarius*	D F
	Great Plains Toad	*Bufo cognatus*	D
	Red-spotted Toad	*Bufo punctatus*	D F
	Woodhouse's Toad	*Bufo woodhousei*	D F M
	Couch's Spadefoot	*Scaphiopus couchi*	D F
	Western Spadefoot	*Scaphiopus hammondi*	D F M
	Canyon Treefrog	*Hyla arenicolor*	F M
	Leopard Frog	*Rana pipiens*	D F M
REPTILES	*turtles*		
	*Desert Tortoise	*Gopherus agassizi*	D F
	Western Box Turtle	*Terrapene ornata*	D F
	Sonora Mud Turtle	*Kinosternon sonoriense*	D F M
	Spiny Softshell	*Trionyx ferox*	D
	lizards		
	Banded Gecko	*Coleonyx variegatus*	D F
	Desert Iguana	*Dipsosaurus dorsalis*	D
	Lesser Earless Lizard	*Holbrookia maculata*	D F
	Greater Earless Lizard	*Holbrookia texana*	D F
	*Zebra-tailed Lizard	*Callisaurus draconoides*	D F
	*Collared Lizard	*Crotaphytus collaris*	D F
	Leopard Lizard	*Crotaphytus wislizenii*	D F
	Short-horned Lizard	*Phrynosoma douglassi*	F M
	Desert Horned Lizard	*Phrynosoma platyrhinos*	D

92

common name	scientific name	habitat
*Regal Horned Lizard	*Phrynosoma solare*	D F
*Side-blotched Lizard	*Uta stansburiana*	D F
*Tree Lizard	*Uta ornata*	D F M
*Desert Spiny Lizard	*Sceloporus magister*	D F
Clark's Spiny Lizard	*Sceloporus clarki*	D F M
Eastern Fence Lizard	*Sceloporus undulatus*	F M
*Western Whiptail	*Cnemidophorus tigris*	D F
Spotted Whiptail	*Cnemidophorus sacki*	D F
Arizona Alligator Lizard	*Gerrhonotus kingi*	M
*Great Plains Skink	*Eumeces obsoletus*	D F
*Gila Monster	*Heloderma suspectum*	D F

snakes

common name	scientific name	habitat
Western Blind Snake	*Leptotyphlops humilis*	D F
Arizona Coral Snake	*Micruroides euryxanthus*	D F M
Regal Ringnecked Snake	*Diadophis regalis*	F M
Western Hognose Snake	*Heterodon nasicus*	D F
Spotted Leaf-nosed Snake	*Phyllorhynchus decurtatus*	D F
Saddled Leaf-nosed Snake	*Phyllorhynchus browni*	D F
Coachwhip	*Masticophis flagellum*	D F
Sonora Whipsnake	*Masticophis bilineatus*	D F M
Western Patch-nosed Snake	*Salvadora hexalepis*	D F
Mountain Patch-nosed Snake	*Salvadora grahamiae*	M
Glossy Snake	*Arizona elegans*	D F
Gopher Snake	*Pituophis catenifer*	D F M
Common Kingsnake	*Lampropeltis getulus*	D F
Sonora Mountain Kingsnake	*Lampropeltis pyromelana*	M
Long-nosed Snake	*Rhinocheilus lecontei*	D F
Black-necked Garter Snake	*Thamnophis cyrtopsis*	F M
Mexican Garter Snake	*Thamnophis eques*	D
Checkered Garter Snake	*Thamnophis marcianus*	D F
Western Ground Snake	*Sonora semiannulata*	D F
Western Shovel-nosed Snake	*Chionactis occipitalis*	D
Banded Sand Snake	*Chilomeniscus cinctus*	D F
Mexican Black-headed Snake	*Tantilla atriceps*	D F
Plains Black-headed Snake	*Tantilla nigriceps*	D F
Sonora Lyre Snake	*Trimorphodon lambda*	D F
Night Snake	*Hypsiglena torquata*	D F M
Western Diamondback Rattlesnake	*Crotalus atrox*	D F
Sidewinder	*Crotalus cerastes*	D
Black-tailed Rattlesnake	*Crotalus molossus*	F M
Tiger Rattlesnake	*Crotalus tigris*	F
Mohave Rattlesnake	*Crotalus scutulatus*	D F
Arizona Black Rattlesnake	*Crotalus viridis cerberus*	F M

Birds of the monument

This checklist is based entirely on written records of observations of birds seen in the monument. Names are in accordance with A.O.U. Checklist of North American Birds, 5th edition, 1957. (See checklist of reptiles and amphibians for description of habitats.)

key to symbols:

D	desert habitat	s	summer resident
F	foothill habitat	w	winter visitor
M	mountain habitat	p	permanent resident
*	species most often seen at lower elevations	t	transient visitor

occurrence	common name	occurrence	common name
D t	White-faced Ibis	D F s	*White-winged Dove
		D p	*Mourning Dove
D F M s	Turkey Vulture	D p	Ground Dove
D p	Black Vulture	D p	Inca Dove
D w	Sharp-shinned Hawk		
D F M p	Cooper's Hawk	D F p	Roadrunner
D F M p	*Red-tailed Hawk		
D F s	Swainson's Hawk	D F M p	Screech Owl
F M s	Zone-tailed Hawk	M p	Whiskered Owl
M t	Ferruginous Hawk	M p	Flammulated Owl
D F s	Harris' Hawk	D F M p	Great Horned Owl
D s	Black Hawk	D s	Elf Owl
F M p	Golden Eagle	M p	Spotted Owl
D w	Marsh Hawk		
D t	Osprey	M s	Whip-poor-will
F M p	Prairie Falcon	D F p	Poor-will
M p	Peregrine Falcon	D F s	Lesser Nighthawk
D F p	*Sparrow Hawk		
		D t	Vaux's Swift
F p	Scaled Quail	D F M s	White-throated Swift
D p	*Gambel's Quail	D F M s	Black-chinned Hummingbird
F M p	Harlequin Quail		
M p	Turkey	D F s	Costa's Hummingbird
		D w	Anna's Hummingbird
D p	Killdeer	M s	Broad-tailed Hummingbird
D F M t	Spotted Sandpiper		
		M s	Rufous Hummingbird
D t	California Gull	M s	Rivoli's Hummingbird
		M s	Blue-throated Hummingbird
F M s	Band-tailed Pigeon		
D p	Rock Dove	D F t	Broad-billed Hummingbird

94

occurrence	common name	occurrence	common name
M p	Red-shafted Flicker	M p	Pigmy Nuthatch
D F p	*Gilded Flicker		
D p	*Gila Woodpecker	M p	Brown Creeper
F M p	Acorn Woodpecker		
D F w	Yellow-bellied Sapsucker	M p	House Wren
F M t	Williamson's Sapsucker	F M p	Bewick's Wren
M p	Hairy Woodpecker	D p	*Cactus Wren
D F p	Ladder-backed Woodpecker	D F M p	Canyon Wren
		D F M p	Rock Wren
F M p	Arizona Woodpecker		
		D p	Mockingbird
D s	Western Kingbird	D t	Bendire's Thrasher
D F s	Cassin's Kingbird	D p	*Curve-billed Thrasher
D F s	Wied's Crested Flycatcher	D F p	Crissal Thrasher
D s	*Ash-throated Flycatcher	M p	Robin
M s	Olivaceous Flycatcher	F M p	Hermit Thrush
D p	Black Phoebe	F M p	Western Bluebird
D p	*Say's Phoebe	D F M w	Mountain Bluebird
M s	Western Flycatcher	F w	Townsend's Solitaire
M s	Buff-breasted Flycatcher		
M s	Coues' Flycatcher	F M p	Blue-gray Gnatcatcher
M s	Western Wood Pewee	D p	Black-tailed Gnatcatcher
D F M t	Olive-sided Flycatcher	D w	*Ruby-crowned Kinglet
D s	Vermilion Flycatcher		
		D F p	*Phainopepla
M s	Violet-green Swallow		
D t	Barn Swallow	D F p	*Loggerhead Shrike
D t	Cliff Swallow		
D s	*Purple Martin	D p	Starling
M p	Steller's Jay	M s	Hutton's Vireo
F M p	Scrub Jay	D s	Bell's Vireo
F M p	Mexican Jay	M s	Solitary Vireo
M p	Common Raven	M s	Warbling Vireo
D F M p	White-necked Raven		
M t	Clark's Nutcracker	M t	Orange-crowned Warbler
		M t	Nashville Warbler
M p	Mountain Chickadee	M s	Virginia's Warbler
F M p	Bridled Titmouse	D s	Lucy's Warbler
D p	*Verdin	M s	Olive Warbler
M p	Common Bushtit	D t	Yellow Warbler
		M s	*Audubon's Warbler
M p	White-breasted Nuthatch	M s	Black-throated Gray Warbler
F t	Red-breasted Nuthatch		

occurrence	common name	occurrence	common name
F M t	Townsend's Warbler	M p	Evening Grosbeak
M t	Hermit Warbler	M t	Cassin's Finch
M s	Grace's Warbler	D F p	*House Finch
D t	MacGillivray's Warbler	F M p	Pine Siskin
D t	Yellowthroat	D F M p	Lesser Goldfinch
D t	Yellow-breasted Chat	D w	Lawrence's Goldfinch
M s	Red-faced Warbler	M p	Red Crossbill
D F M t	*Wilson's Warbler	D F w	Green-tailed Towhee
M s	Painted Redstart	F M p	Rufous-sided Towhee
		D p	*Brown Towhee
D p	House Sparrow	D t	Lark Bunting
		D F w	Savannah Sparrow
D t	Meadowlark sp.	D w	Vesper Sparrow
D s	*Hooded Oriole	D t	Lark Sparrow
D F s	Scott's Oriole	D F p	Rufous-winged Sparrow
D s	Bullock's Oriole	F p	Rufous-crowned Sparrow
D w	Brewer's Blackbird	D F p	*Black-throated Sparrow
D F s	Brown-headed Cowbird	F w	Oregon Junco
D s	Bronzed Cowbird	F w	Gray-headed Junco
		M p	Mexican Junco
D F M s	Western Tanager	F w	Clay-colored Sparrow
M s	Hepatic Tanager	F w	Chipping Sparrow
		D F w	Brewer's Sparrow
D p	Cardinal	F p	Black-chinned Sparrow
D p	Pyrrhuloxia	D w	White-crowned Sparrow
M s	Black-headed Grosbeak	D·F w	Fox Sparrow
		D F w	Lincoln's Sparrow
D w	Lazuli Bunting		

Mammals of the monument

This checklist names mammals that have been observed, and those that, mainly according to collection records and distribution maps in *The Recent Mammals of Arizona*, by E. Lendell Cockrum, 1960, may occur in the monument. One species, the desert kangaroo rat, probably occurs only in the Tucson Mountain Section. (See checklist of reptiles and amphibians for description of habitats.)

key to symbols:

D	desert habitat	M	mountain habitat
F	foothill habitat	*	species most often seen by visitors

habitat	common name	habitat	common name
M	Vagrant Shrew	D F	*Desert Cottontail
D F	Gray Shrew		
		D F M	Rock Squirrel
D F	Leafnose Bat	D F	*Yuma Antelope Ground Squirrel
D F	Hognose Bat		
D F	Longnose Bat	D	*Roundtail Ground Squirrel
D F	Yuma Myotis	F M	Cliff Chipmunk
D F	Cave Myotis	M	Tassel-eared Squirrel
M	Long-eared Myotis	M	Arizona Gray Squirrel
F M	Fringed Myotis		
F M	Long-legged Myotis	D F M	Valley Pocket Gopher
D F M	California Myotis		
F M	Small-footed Myotis	D F	Silky Pocket Mouse
F M	Silver-haired Bat	D F	Arizona Pocket Mouse
D F M	Western Pipistrel	D F	Hispid Pocket Mouse
D F M	Big Brown Bat	D	Desert Pocket Mouse
D F	Red Bat	D F	Rock Pocket Mouse
D F M	Hoary Bat	D	Bannertail Kangaroo Rat
D F	Western Yellow Bat	D	Merriam Kangaroo Rat
D	Spotted Bat	D	Ord Kangaroo Rat
D F M	Western Big-eared Bat	D	Desert Kangaroo Rat
D F	Pallid Bat	D F	Western Harvest Mouse
		D F	Fulvous Harvest Mouse
D F M	Mexican Freetail Bat	D	Cactus Mouse
D F	Pocketed Freetail Bat	D	Merriam Mouse
D F	Big Freetail Bat	D F	Deer Mouse
D F	Western Mastiff Bat	D F	White-footed Mouse
		F M	Brush Mouse
D	Antelope Jackrabbit	D F	Southern Grasshopper Mouse
D F	*Blacktail Jackrabbit		
M	Eastern Cottontail	D F	Hispid Cotton Rat

97

habitat	common name	habitat	common name
D F	Whitethroat Woodrat	F	Striped Skunk
M	Mexican Woodrat	D F	Hooded Skunk
		D F	Hognose Skunk
D F M	Porcupine		
		F M	Jaguar
D F M	*Coyote	D	Jaguarundi
D F	Gray Wolf	F M	Cougar
D	Kit Fox	D F M	Bobcat
D F M	Gray Fox		
M	Black Bear	D F	*Peccary (Javelina)
D F	Ringtail	D F	*Mule Deer
D F M	Raccoon	M	Whitetail Deer
M	Coati		
		D F	Bighorn
D F	Badger		
D F M	Spotted Skunk		

www.ingramcontent.com/pod-product-compliance
Lightning Source LLC
Chambersburg PA
CBHW051337170526
45166CB00002B/851